SCANNER
MODIFICATIONS
AND ANTENNAS

JERRY PICKARD

PALADIN PRESS · BOULDER, COLORADO

Scanner Modifications and Antennas
by Jerry Pickard

Copyright © 1999 by Jerry Pickard

ISBN 1-58160-041-0
Printed in the United States of America

Published by Paladin Press, a division of
Paladin Enterprises, Inc.
Gunbarrel Tech Center
7077 Winchester Circle
Boulder, Colorado 80301 USA
+1.303.443.7250

Direct inquiries and/or orders to the above address.

Visit our Web site at www.paladin-press.com

CONTENTS

DEDICATION

Like any book, this one is the product of many people.
Thanks to all the knowns and unknowns who helped
and gave support.

Special thanks, of course, to my wife, Barbara
Who put up with nearly everything,
and to Laura Quarantiello whose shoulder I cried on,
and particularly to Jaime Jay Carver, who with patience and
sometimes true grit,
guided me and the computer through much frustration.

Warning

This book is written *for information purposes only*. Neither
the author, publisher, nor seller of this book assumes, or
will assume, any responsibility for the use or misuse of the
information contained herein, nor will they be responsible
for the consequences thereof, either civil or criminal. If you
have doubts or questions about what is or is not unlawful,
please consult an attorney.

Foreword

It's a dangerous person who backs off the screws on the cover of a scanner and opens it to light and view and hot soldering iron. Somewhere, months or years ago, this scanner sat open to the world on a factory bench. A company technician checked it, tagged it, and closed it like a surgeon pulling tight the last stitch. It traveled down the conveyor belt, was slipped into its plastic packaging and on into the cardboard box that would be its home for months or years. . . until somewhere an owner bought it, took it home, broke the seals, and gave it a place to work, a shack of its own.

Its cover remained in place, forever sealed by the factory technician, perhaps thousands of miles away by now working on some other unit. The scanner's heart and soul remained in a darkened space, asked to work in silence and blackness, with only the breath of a circulating breeze to cool it. Until a dangerous man, or maybe a dangerous woman, approached, backed off the screws on the cover and opened it. Exposed the heart and soul of it. Probed and poked it with tweezers, a screwdriver, a smoldering soldering iron. Dangerous fingers began to alter this scanner, this factory-assembled, factory-sealed, factory-approved piece of electronics. Fingers that worked a sort of high-tech hocus-pocus. Later, when the fingers were withdrawn, the cover was back in place and power was applied, this factory-common scanner, like a thousand others that had rolled down the line, was very suddenly not like a thousand others. It was uncommon. It was different. It did things those thousand others couldn't possibly aspire to in their cloned little lives. **It had been modified.**

One of my scanners was a stock PRO-2006 until my dangerous, dancing fingers popped the cover and went hacking into its innards. Like many others, I knew that the scanner I now owned came off the ways with the potential for much more within it. I knew that, with a little knowledge and the right tools, I could improve the performance of my scanner and thereby improve my ability to scan the airwaves. Oh, I put it off, of course. I convinced myself that modifications were dangerous and unnecessary and the sole realm of the professional electrician. But, inevitably, like a bottle on a beach or a genie in a lamp, I had to release what I knew was inside waiting. I had to take the chance.

I was smart enough, or maybe just lucky enough, to find a pro to advise me before I did the deeds that would forever alter my machine. Though the information on modifications could be found if you looked hard enough, it still lived an underground existence. It was still heap big magic to crack the case and diddle the diodes. Someone who did was very much like a computer hacker and frowned on just the same. I was given the right advice and the right instructions. The modification worked. My scanner was forever and eternally a busted-warranty, hyped-up, turbo-technical machine. But, I kept quiet about it. Folks would ask, but I'd shake my head. "Sorry, can't help you. I've never done a mod." I lied. I avoided the subject. I couldn't admit to scanner hacking. And then I found Jerry Pickard.

He didn't lie, he didn't avoid the subject. Jerry had modified scanners and he wasn't afraid to say so. As a matter of fact, he had done the mods himself. Finally, a kindred spirit! He acknowledged clearly that modifications would void your factory warranty, that you might clip the wrong diode, remove 4712 instead of 4713. You do it all at your own risk, he said. Do it wrong and repairs will cost you, do it right and you'll have a scanner that does everything it can really do, everything it was meant to do . . . and I'll tell you how. There were other gurus like Jerry out there who were writing books on modifications and giving advice on scanner upgrades. But their instructions came with a prerequisite: you had to know your IC from your AC. If you didn't, you were doomed to sit with an open scanner under your hands, powerless to find the right part to perform the modification. You'd never get past "remove all four Phillips-head screws," which was perhaps the only understandable part of the explanation.

Jerry changed all that. He approaches modifications from the bottom up and uses words you don't have to look up, diagrams you don't need an electrical degree to understand, and step by step procedures that are clear and concise. Best of all, if you don't understand something, he isn't afraid to talk you through it. He spends time on the telephone and on the computer helping people. He never asks for anything in return and he usually never gets it. Here is a man who has made scanner modifications the realm of the common user, instead of the realm of the "dangerous man." There are only so many times you can answer "tell me the mod for. . . " Someone finally convinced Jerry to write it all down in one place and that's why this book exists. It will improve your equipment and help you to get the most from your scanner. This book is for you and me, novice and experienced user alike. Because somewhere out there is a brand new scanner, waiting for a shack to call its own, waiting to be opened like Pandora's Box.

>Laura Quarantiello
>Author of:
>*The Citizen's Guide to Scanning*
>and
>*Cop Talk! Monitoring Law Enforcement Communications*
>and
>*Air-Waves—The Aviation Monitor's Handbook*

1

Scanner Laws

Everyone who owns a scanner has a responsibility to use it in a manner that represents all the other people involved in the scanner "hobby." In my opinion, this means not interfering in any way with the services we are able to monitor. Chasing crimes or fires can be fun, but getting in the way does not serve well. We have, with the scanner, a special opportunity to see the way our tax dollars are spent, listening to people just like ourselves going into harm's way, risking (and suffering) at least stress, and frequently personal injury.

We are curious about the world around us, and the scanner offers an unique view into that world that many others do not have.

Overwhelming data shows that scanner users often provide information in times of disaster. A woman in Florida kept a building full of seniors appraised of a coming hurricane when no other source of information was available. It seems a small thing, perhaps, but if I were a terrified person in the building, I'd like to know the reality of my surroundings. I have files full of stories about citizens who provided information about impending crimes, and about those who helped law enforcement officials in the location and apprehension of criminals on every level. Many handicapped people have contacted me, saying that their scanner provides them with company, and allows them to know the things that are happening around them.

Regrettably, a scanner can also be used to "invade" privacy. Far too many of the population fail to realize that when they pick up a cordless or cellular phone, they are broadcasting to anyone with the capability of receiving that broadcast. The media has made it seem that anyone with a scanner has little else to do but listen in to the private and important secrets of the guy down the street.

FEDERAL LAWS

In my opinion, the original law concerning communication monitoring, the Communication Privacy Act (CPA), written in 1934, was necessary, complete, and served everyone, broadcaster and monitor alike in the best possible way. Briefly, it allowed both broadcasters and monitors free access to the radio spectrum, but provided penalties for misuse. Considering that radio (and other) waves are all-pervasive and penetrate our whole environment, the provisions set out in the CPA made excellent sense. Any communicator who required total privacy had (and has) easy access to scrambling and encryption devices, as well as the fundamental awareness that anything broadcast over the radio spectrum could be heard with an appropriate receiver.

With the successful advent of cellular telephones, however, some unfortunate changes were made. Rather than choosing an encryption method to insure privacy for Cellular Mobile Telephones (CMT's), the cellular industry, through Congress, was able to pass laws restricting monitoring of "their" part of the spectrum. In 1986, the Electronic Communication Privacy Act (ECPA) was passed. Suddenly it was illegal to listen to any communication in the small part of the radio spectrum allotted to the cellular phone bands. Scanner manufacturers were forced to block availability to that part of the radio band, and I imagine it cost them megabucks to alter radios then available, and more to maintain adherence to the new laws. Oddly, only scanning receivers were affected. If your radio could be tuned continuously, your access to that part of the band was not restricted. In fact it was (and still is) possible to hear those calls on the high end of almost any TV receiver (tune around channel 79). (Note: The complete law can be looked up at almost any library. See H.R.-4952, Public Law 99-508, Amendment to Title 18.)

Now there are new limits further restricting scanner manufacturers. To comply with the Telephone Disclosure and Dispute Resolution Act (Public Law 102-556), after April 26, 1993 no scanner could be certified that was able to cover, or that could be made to cover, the radio spectrum between 824.000 to 849.000 MHz and the spectrum between 869.000 and 894.000 MHz. The amendment follows:

Section 302 of the Communications Act of 1934 (47 U.S.C. 302). Under Section 403 ("Interception of Cellular Telecommunications")
(a) Amendment- Section 302 is amended by adding at the end of the new subsection:
"(d) (1) Within 180 days after the date of enactment of this subsection, the Commission shall prescribe and make effective regulations denying equipment authorization (under Part 15 of Title 47, Code of Federal Regulations, or any other part of that Title) for any scanning receiver that is capable of —
"(A) receiving transmissions in the frequencies allocated to the domestic cellular radio telecommunications service,
"(B) Readily being altered by the user to receive transmissions in such frequencies, or
"(C) being equipped with decoders that convert digital cellular transmissions to analog voice audio.
"(2) Beginning one year after the date of the regulations adopted pursuant to paragraph (1), [see below] no receiver having the capabilities described in subparagraph (A) (B) or (C) of paragraph (1) as such capabilities are defined in such regulations shall be manufactured in the United States, or imported for use in the United States."

Here's the part of section (1) that is pertinent :

> Sec. 402 Technical Amendments:
> "(C) May, by rule or order, exempt from the requirements of paragraphs (1)(A) (iii) of this subsection calls to a telephone number assigned to a cellular telephone service that are not charged to the called party, subject to such conditions as the Commission may prescribe as necessary in the interest of the privacy rights this section is intended to protect."

Phew! All that forbidding, dense verbiage to deprive scanner users of a tiny section of the spectrum that could readily be denied by simple changes in the cellular phone, and the cost passed on to the cellular phone consumer!

I believe in personal privacy, and readily accept the idea that eavesdropping on someone's private conversations is, if nothing else, an indication of low morals. I do not believe that the cellular phone sellers should use restrictions of other's equipment to fallaciously promise privacy to a few. Removing a part of the open-air spectrum for that purpose seems to be a greater violation of *everyone's* freedom. As someone wrote to me, "if they don't want me to hear their broadcasts, keep 'em out of my backyard."

Just to note, the regulations don't address *personal* modifications which as of this writing are perfectly legal to do and possess, as long as you are not a manufacturer.

*[**Publisher's Note:** In February, 1997, the FCC made a startling announcement that it was illegal to modify a scanner for cellular reception. See page 165 for the text of the directive. It appears to be aimed at for-profit modifiers, not at individuals working on their own scanners, but who knows what Big Brother might do. You have been warned!]*

STATE AND CITY LAWS

States may have specific laws regarding scanner use; it is up to the scanner user to be aware of them. In California for example, it is illegal to monitor cordless phones and baby monitors.

To add to the confusion, some cities have ordinances to restrict scanner use, even though the state might not. There is at least one state where it is not legal to have a scanner at all, some states have various restrictions on mobile scanner use, and most states add penalties if a scanner is used in a crime. Almost all states have statutes against interfering with the operations of the utilities (police, fire, etc.). Since the local laws are subject to change, I won't try to list the laws I know of currently, since when you buy this book, many of the laws might have changed. Knowing the law is up to you, and if you are a traveler, learn how the law works in each area along your route. In some you'll find the need for permits, others have no restrictions, and in others, there are absolute prohibitions.

ABOUT CELLULAR PHONES

The Communications Act of 1934 seems reasonable, providing certain penalties for misusing intercepted communications. At that time, the reality of a radio wave's penetration into the atmosphere was a given, and the public was given the responsibility of use. This worked until 1986, at which point the Cellular Mobile Phone lobby entered the picture. In order to promote the myth of privacy, the lobby persuaded Congress that it should make new laws restricting those parts of the electromagnetic spectrum that CMT (Cellular Mobile Telephone) controlled. Manufacturers at that point were required by law to change the circuitry of scanners to conform to those restrictions. This was done primarily by blocking those parts of the scanner's circuitry covering that area of the scanner's spectrum. The way the law was written, it allowed a scanner owner to modify his or her radio to receive the CMT frequencies, but made it illegal to hear it!

"Reasonable expectation of privacy" was the goal of the cell-phone lobby to allow the CMT manufacturers the erroneous statement that cellular phones were as private as hard-wired phones. As the public became aware that this was not so, the cell-phone lobby again manipulated Congress to further restrict a scanner's capability to receive the calls within those specific frequencies.

2

What's A Mod?

Once in a while, I'll get a letter stating, "I want all the mods for so-and-so scanner, and by the way, what is a mod?" This book is written for those inquirers, and for you.

In every case possible, I've written in as simple a method as possible so just about anyone who can determine the hot end of a soldering iron, or clip his/her toenails can do the modifications in this book.

My feeling is that anything that enhances the operation of a scanner falls into that category, so you'll find things here on antennas, circuits, tricks I've learned from experts, and odds and ends that often came from pure serendipity. In some cases the possibility of breaking the law exists if this information is misused, but the modifications themselves are perfectly legal to do. It is my expectation that anyone using this book will do so responsibly.

Scanning is an excellent hobby, often allowing the scanner user an insight into a world where individuals just like us put themselves in harm's way for our protection. For example, scanners can show how our tax money is spent, open other worlds, some funny, some exciting, some grim, some educational, but never uninteresting.

I think some of the people who choose to modify their radios do so for the same reason Hillary climbed Everest, "Because it's there." Whatever your reasons, I hope this book is a useful guide.

No assurance can be given for the results gained from this book, since it is not possible to evaluate a person's skills or experience. The reader must be aware of the likelihood of negating a scanner's warranty, and doing damage to your scanner is always a possibility. While I've done most of these modifications in my own shop, not all have been tested, and neither the publisher, others involved, or myself take any responsibility for the results of these modifications.

I strongly urge anyone doing scanner modifications to have the service manual for a given radio. Reading schematics isn't that hard, and knowing more about your scanner can't hurt. Where possible, I've chosen to use graphic drawings to clarify the work. In other words, instead of using electronic symbols, the parts are drawn out.

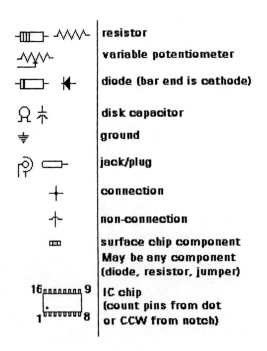

resistor

variable potentiometer

diode (bar end is cathode)

disk capacitor

ground

jack/plug

connection

non-connection

surface chip component
May be any component
(diode, resistor, jumper)

IC chip
(count pins from dot
or CCW from notch)

3

Tools And Tricks

Working inside your scanner, especially a hand-held, can look forbidding. With surface-mount chips and limited space, some adjustment has to be made for "flea size" work. Don't panic, it's really easier than it looks, but some care is obviously required.

First, clear a work space without any clutter other than just the tools and materials you need. A tray or two for parts, both removed and added, is very helpful. One of the things I sometimes do is to stick a strip of double face tape to a piece of cardboard, to hold parts in order when I'm not sure of a sequence of disassembly, so I can see the order of parts when it comes time to put the thing back together. It helps when you lose track of which screw came from where.

Static insurance: There are components in scanners that can be destroyed by the tiny spark created from rubbing your shoes on the floor, or even from a fan blowing on you and your work area! There's a simple way to prevent this from happening. You can make an anti-static wrist strap from a couple of feet of ordinary swivel chain, available from any hardware store, by putting a loop at one end for your wrist, and a 100K resistor at the other end, connected to the ground screw at a handy electrical socket. (The screw that holds the cover on the socket will do nicely.)

Soldering tools: If your soldering "iron" is just too big for the job, or too hot, wrap a piece of bare copper wire tightly around the tip, leaving the end extended and do the job with that. Have a wet sponge or rag at hand to keep your soldering tip very clean. A lot of bad joints or slow soldering can come from a dirty tip.

Holes: For plastic cases, I have a semi-retired small soldering pencil that's easy and quick to use. To size the final hole, a tapered reamer does a nice job, with a little clean-up with a knife or file.

For metal cases (chassis) I never drill a hole in the case larger than 3/16 inch, and size the hole with a tapered reamer. The local auto parts store has been my source for appropriate sized rubber tubing to make drill stops to prevent drilling into the "danger zone." Usually, only about 1/8 inch of the drill bit needs to be exposed. Locate your holes carefully, using a center punch.

Often, I'll use a magnet in the drill area to catch metal particles, and then a brush and shaking to make sure all the "foreign objects" are removed. It only takes a minute particle of conductive material to cause havoc in your precious scanner.

Wire: For jumper wire, try to find a length of multi-conductor wire of the type used to multiplex phone systems. It's about 20 gauge, stranded, and nice and flexible. I watch the thrift stores for this, and it costs next to nothing. On the occasions when I've needed to "undo" a mod, or try components, wrapping wire does the job. It's fine enough to easily locate on a tiny soldering pad, and small enough to bring out of an assembled case for external testing. Trying different crystals, for example, 30 ga, 50 ft, Radio Shack #278-501.

A 1N914 diode will replace the diodes used for blocking and switching in scanners.

Above all, remove the power source from your scanner when doing any work. Consider the obvious shock hazard, as well as the strong possibility of damage to your scanner.

4

Sensitive Components

POP! Although you might never hear it, that's the sound of carelessness when you're working with microchips and transistors. Under high magnification, you can see some amazing craters and "meltdowns" that are caused by a tiny spark you probably are unaware of. Elsewhere in this book, I've mentioned using static straps and static work pads so there's no need to labor that point again, except to remind you not to be careless with your work, ever.

There are a couple of circuits in this book that use transistors, and if you are a beginner at this stuff there are a couple of things you should know. Transistors don't like excessive heat. If you put your hot gun on an unprotected transistor for any length of time (more than three seconds), the result will be the same as if you did the same thing to a bug. The poor little thing will just curl up its tiny legs and join that great electronic graveyard in the sky!

Take heart, my friend, there are ways to protect your transistor and other heat sensitive parts from "the evil that men do."

Heat sinks: Anything that will conduct heat away from the transistor will do the job. There are clamp-like devices available in any electronics store that you can simply clip on the lead you're soldering between the connect point and the component.

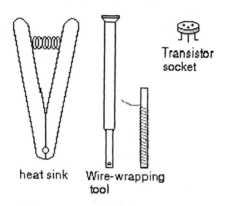

Transistor socket

heat sink   Wire-wrapping tool

Wire wrapping: This is an elegant solution that can, if you decide, be used cold to make a reliable connection, and if you touch the connection with a bit of solder the connection is absolutely solid. To make the connection, a piece of very fine wire (30 gauge) is stripped and inserted in the "mouth" of the tool and out the side through a built-in groove. Then the tool is placed over the pin of a wire-wrap socket, or simply over a component's lead, and rotated until the connection is made. There are wire-wrap sockets available to fit just about any component a hobbyist might need. Both wire-wrap sockets and transistor sockets are made to fit easily available pre-punched perf-board (also called vector board).

Transistor sockets: Another "handy helper" to save some of the grief that can befall the electronic hobbyist. The leads on a circuit can be soldered to the socket without "heat frustration" and when the hot part is over, the transistor's leads are clipped to length and inserted into the socket. Easy, huh?

NOTE: Here's an easy, dumb way to remember what kind of transistor you're working with. Look at the arrow in the schematic and see which way it points. There are two kinds of transistor, determined by polarity, PNP and NPN. Notice whether the arrow is **N**ot **P**ointing i**N** = NPN.

Perf-board: This is probably the best thing to happen for the electronic hobby builder. All it is is a non-conducting material perforated with holes sized and spaced for electronic hobby use. (There are different configurations of perf-board, made to mount specific components such as chips, with the copper traces already in place. Explore!) Perf-board makes laying out a circuit as easy as possible, and will give a head start to any project.

Short cut to cutting perf-board: With a fine-toothed hobby razor saw, just score the surface of the perf-board along the line you want the cut to be, and you will find you can snap it very accurately. The perforations make great guides, and scoring the line along the holes makes the job simple. Often, to keep a project small (part of the fun!) I'll place the components first, wire things up, and then cut the board. Like sculpting an elephant. (Take a big block of stone and then cut off anything that doesn't look like an elephant.)

If you are a beginner, don't let inexperience hold you back. All of us started without any experience, made mistakes along the way (and still do!), but whatever frustrations you suffer, the end result is always worth the price. It feels good to point at a project or a mod and say "I made that!"

5

Cellular Programming PRO-series Scanners

Since there are 833 frequencies allotted for the cellular phone bands, it isn't practical to try to enter separate frequencies into your scanner. The best way to monitor the cell-phone band is by using a dedicated search bank. Not only is it easier, by "setting aside" the cell band you keep the regular channels open for more practical use.

TO PROGRAM THE PRO-2004/5/6:

| program | (1) | limit | (Lo) 868.980 | limit | (Hi) 891.990 | enter | △ |

Press **PROGRAM**

Press **1 : LIMIT :** (the scanner display will change to LO)

Enter **868.950** then press **ENTER**

Press **LIMIT:** (the display will change to Hi)

Enter **894.000**, then press **ENTER**:

Then press the **UP ARROW** (under **LIMIT** on your keyboard)

Your scanner should begin to scan and receive CMT calls.
To cancel the CMT calls, just press **SCAN**
To restart CMT, Press **PROGRAM : 1 : LIMIT : UP ARROW**

PRO-series scanners will automatically default to 30 kHz spacing necessary to receive CMT calls correctly. Not all other brands of scanners have this 30 kHz default.

NOTE: I've only seen it once, with a PRO-2006, but if the scanner fails to default, press **RESET**. If you simply enter 868.950 lo/894.000 hi, usually the PRO-series scanner will default to the exact frequencies!

6

PRO-2004/5/6 Modifications

PRO-2004: Cellular and 400 Channels

It's worth having one of these "old soldiers" in the pristine state just so you can do the mods! With very little effort, you can turn this one into quite a different scanner.

Disconnect the power and antenna, remove the bottom cover, and locate "the box" marked PCB-3. Inside you'll find a diode matrix, and all you need to do to restore cellular capability is to snip or remove diode D-13. NOTE: if yours is a really old one and you don't find this diode, it's OK. For some reason, a run of these scanners were made with that diode on the underside of the board; all by itself, too - easy to find.

While the case is open, at the pads for D-510, solder in a 1N914 diode, and you add another 100 channels!

This particular radio is a beauty in my opinion, with all the range of the later PRO-2006.

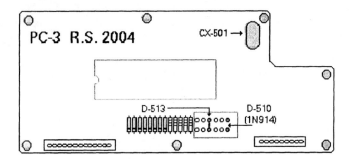

PR0-2004 SQUELCH MODIFICATION

If you find the squelch on this scanner too "loose" for your taste, there is a way to tighten it up (narrow the range) and make it adjustable to suit. Remove the power and antenna, and open the case. Locate chip IC-2 and alongside it find R-148. Clip and separate the lead at the top of this resistor, making sure you have enough lead to add a 200 K potentiometer in series as shown. If you need to, it's OK to add a bit of hook-up wire. Now you can limit the squelch range by adjusting the pot.

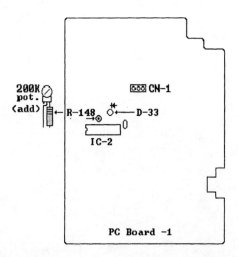

PR0-2004 HYPERSPEED MODIFICATION

To almost double the speed of this scanner is fairly easy, but you need to get to the underside of board PC-3 to find the ceramic oscillator (CX-501). Since this runs at 7.37 MHz, a 10 or 12 MHz crystal will replace it for a significant speed increase (80% or so). Too high a value will cause the keyboard to lock up. See the other hyperspeed mods for cautions on handling crystals. Remember, speeding the scan rate shortens the delay period.

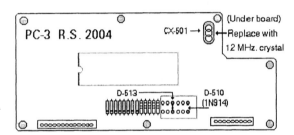

PRO-2005 CELLULAR MODIFICATION

The modification for the -2005 is slightly different from the PRO-2006. Clipping D-502 restores CMT, but D-503 appears to have no effect on the scan speed. For a speed increase, other than changing the ceramic resonator, it is necessary to install a 1N914 diode at D-501. This will add about 25% to your speed with no other effect. Pay attention to the polarity of the diode; it is the same as the others in the matrix. Although the position of the diodes may seem peculiar (see inset) they are correct. You might find it necessary to remove the front panel to be able to do this modification.

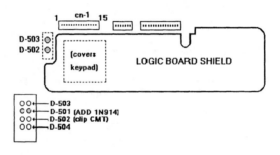

PRO-2005 HYPERSPEED MODIFICATION

Nearly any scanner's speed can be increased considerably, and this one is no exception. There is a price to pay, though. As the scan rate increases there is a proportionate decrease in the scanner's delay time. See the front panel removal directions for the PRO-2006. Locate CX-501 (ceramic resonator). Replace it with a 10 MHz crystal. When you remove CX-501, you'll see that it has three pins. Solder the two-pin crystal using the two outside pads. NOTE: Even with power removed, it's possible to have a chaotic accident by shorting either outside pad to ground or other components. Use care, and insulate the crystal with tape or spaghetti tubing. Remember also, crystals are shock sensitive, so be gentle as you cut or bend the leads. If you hold the lead with a pair of needle-nose pliers against the crystal and do your cutting and bending on the side away from the crystal, it's much safer.

RS 2005/6 logic board (top view)

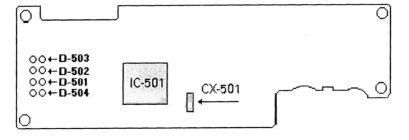

PRO-2005 SIMPLE "S" METER

If you like to see moving meters (I do!), this is a worthwhile project, and very accurate. I use mine mostly for antenna comparison, using a given station with a known signal. Mounting the meter is left up to you, since the variety of meters and housings is endless. The meter I use is an old 0 to 50 microamp meter from Micronta and gives excellent results.

The circuit is easy to build, just point to point on a small piece of perf-board.

Parts list:
2   Capacitors, 0.01 µF / 50 wvdc.
2   Diodes, 1N34A.
1   Resistor, 33-k, 1/4 watt
1   Meter, 0 to 50 millivolts
1   Potentiometer, 1 to 10 K ohms.
Appropriate plugs and jacks to suit.

The circuit board can be installed inside the scanner, or inside the meter housing. The pot is used to set a full range on the meter from a constant strong station. (I live near enough to a NOAA weather station to use that.) This circuit can also be used with an "S" meter salvaged from a CB set, or directly with a digital voltmeter.

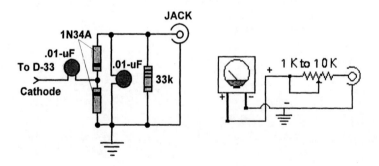

PRO-2004/5 SHORTWAVE CONNECTION

Just by adding a couple of components, you can inter-connect a communications receiver to your 2004/5 scanner. Depending on the receiver, this will allow you to hear single sideband and other signals that your scanner normally can't detect! This modification (depending on your shortwave receiver) can also allow you to add noise blanking, AM narrow and wide receive, tone control, fine tuning, and other features. It works by taking the raw signal from your scanner's 3rd IF stage to the shortwave receiver and uses that to process the signal. Whatever features are available to your SW set then can be used to manipulate your scanner's output...

Careful shielding is needed for this circuit to work at it's best, so use high quality coax inside your scanner, and for the jumper to the shortwave receiver. Very close to the center of the 4004/5's linear (main) board, you'll find TP (test point) 5, which is the 455 KHz output point. It is right alongside diode D-31.

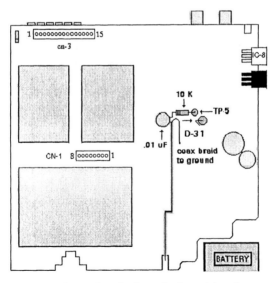

You won't need a circuit board for this, just connect the components as shown in the drawing using leads cut as short as possible. Ground both ends of the internal coax's braid at convenient points. Being very careful, drill a hole in the back of the scanner, and install a BNC connector. This is also a very good time and place to add a ground lug to your scanner for this and other modifications. Make up a jumper as long as you need (but no longer) to reach from the scanner to the shortwave receiver, with an appropriate connector at each end. NOTE: The PRO-2005 does not have test point 5, but the cathode of diode D-31 is the same point.

PRO-2006 CELLULAR MODIFICATION

This may be the most wanted modification, and certainly, it's one of the easiest. At the back of the cover, remove the two top Phillips-head screws, and slightly pull the cover towards you, releasing it. Lift the cover, and you'll see the speaker wires. It's easy to unplug the speaker connector from the circuit board, but there's enough slack (just) to work with the speaker plug in place. Set the cover out of your way. With the back of the scanner facing you, look forward to the metal shield covering the inside of the keyboard (the logic board in this scanner).

Through the cut-out in the upper left corner of the metal shield, you'll see two glass diodes, D-502 and D-503. With a cutter (cuticle scissors work nice for this) snip one lead of each of these diodes and separate the cut ends slightly to prevent re-contact. The cutting of both diodes re-opens the blocked CMT bands, and also slightly increases the scan rate, from 26 to 30 channels per second. With care, reassemble the scanner.

PROGRAMMING: Since there are 833 CMT channels in use, the most practical way I've found to "test" is to use a dedicated search channel and keep it that way for future "testing."

PRO-2006 HEADPHONE AUDIO MOD

Probably to protect our shell-like ears, many scanners have attenuated the audio at the headphone jack. Here's an easy modification to bring the audio up to the speaker level, so the first time you test it, be aware! Remove the power cord from the scanner. Within the wire harnesses that run across the top of the 2006, you will find three twisted wires, black, white, and gray. At a point you find convenient, separate the wires enough to strip a small section of the black wire. It isn't necessary or prudent to cut through the wire, just bare enough to add a short piece of hook-up wire long enough to work with. Solder and insulate. Near the memory battery locate the point on the printed circuit board where the red and black wires from the battery box connect to the board. Trim most of the lead from a 10K ohm, 1/8 watt resistor to the pad where the black wire from the battery box is connected. Solder together the other lead from the resistor and the hook-up wire connected to the black wire from the harness and the deed is did!

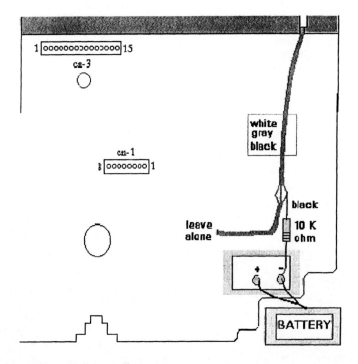

PRO-2006 80% SPEED INCREASE

Now that many public services, and others, have moved up to 800 MHz trunk systems, a faster scanner is even more useful. This mod is a bit more complex than others in this book, and with the speed increase, your delay time will be shortened. The most difficult part of this modification is not the electronics, but the need to remove the scanner's front panel to access the logic board.

RS-2006 LOGIC BOARD (top view)

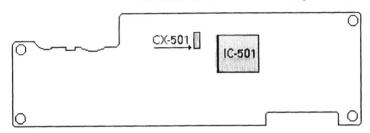

First, disconnect the power and antenna. Carefully pull off the squelch and volume control knobs, and take off the scanner's top and bottom covers. Disconnect all the cable groups from the circuit board, except the CN-3 connector (15 pin). Disconnecting this one will empty your scanner's memory. Unsolder the ground straps from the underside of the main board. Locate and remove the four screws (two on each side) that hold the front panel to the chassis. Pull the front panel free. There is a ground wire near the volume control that goes to the RF shield. Desolder it from the shield. Locate and remove the six screws that hold the logic board to the front panel. and find connector CN-501 (13 pin). There are no wires on this connector. Slip a knife blade under this connector and gently pry it up until it, and the logic board, come loose. Take a break!

Unsolder the points that hold the RF shield to the logic board, and approximately at the center of the board, find the ceramic resonator (CX-501). It can be removed from the top side of the board with patience. Although it is a three-lead component, the center lead can be disregarded. Replace it with a 10 MHz crystal, using the two outside pads where the ceramic resonator was located. Clip the leads on the crystal to the proper length, and bend them at a right angle to fit the spacing of the pads. By working carefully and pre-tinning the leads on the crystal, it can be installed from the top side of the board. Re-assemble the radio carefully, remembering to reconnect all the grounds and plugs.

PRO-2006 SIMPLE "S" METER

If you like to see moving meters (I do!), this is a worthwhile project, and very accurate. I use mine mostly for antenna comparison, using a given station with a known signal. Mounting the meter is left up to you, since the variety of meters and housings are endless. The meter I use is an old 0 to 50 microamp meter from Micronta and gives excellent results.

The circuit is easy to build, just point to point on a small piece of perf-board.

> ***Parts list:***
> 2 Capacitors, 0.01 μF / 50 wvdc.
> 2 Diodes, 1N34A.
> 1 Resistor, 33-k, 1/4 watt
> 1 Meter, 0 to 50 microvolts
> 1 Potentiometer, 1 to 10 K ohms.
> Appropriate plugs and jacks to suit.

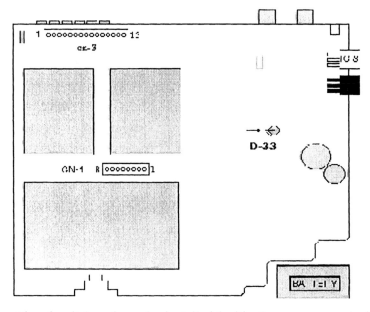

The circuit board can be installed inside the scanner, or inside the meter housing. The pot is used to set a full range on the meter from a constant strong station. (I live near enough to a NOAA weather station to use that.) This circuit can also be used with an "S" meter salvaged from a CB set, or directly with a digital voltmeter.

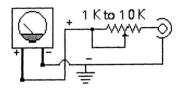

Roughly in the center of the main circuit board, find diode D-33. The cathode end is closest to the board, and that's where you need to attach a piece of hook-up wire long enough to route to the new circuit board. The ground side can go to the case in any convenient place (I drilled the case and installed a ground lug).

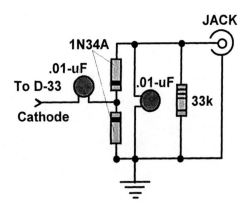

PRO-2006 CENTER ANALOGUE METER

Confession! I have a weakness for meters and flashing lights. This meter is useful on a practical level, if only to see if the transmitter of a given station has drifted (a few have, in my area), but the real fact is that I have an economy-sized old Triplett that I had to use somehow and here's how.

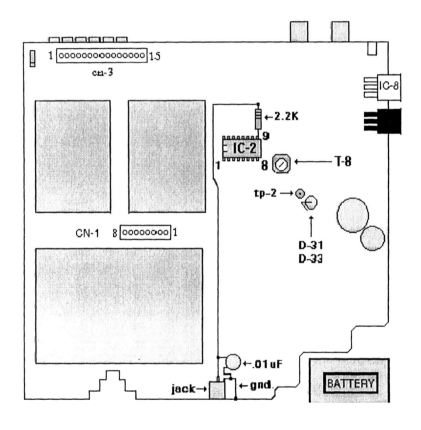

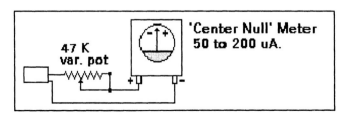

Find pin 9 of IC-2 as shown and either do the work there or locate TP (test point) 2 nearby on the board. This will give you the output you need. To either point, solder a 2.2 ohm resistor (with most of the lead cut off) and to the other end solder enough hook-up wire to reach the rear of the chassis, where you'll install the jack of your choice, or a grommeted hole, to carry the leads to the meter. Near the chassis frame, solder in a .01 μF capacitor from the "hot" lead to ground.

HIT COUNTER FOR PRO 2004/5/6

Wonder how many calls your local law guys took last night? Or the number of transmissions per hour your airport's tower handles? Or whether that "hot" frequency your friend gave you ever works? For 20 bucks or so there's a way to find out, and also to track any other channel's usage.

Radio Shack sells a device called the LCD ELECTRONIC COUNTING MODULE, Part #277-302. It's a counter that advances by 1 each time an add-on switch is closed. It lends itself well to operating with a scanner. Stated most simply, in this configuration, the module switch is activated by the scanner's squelch.

Wiring is simple. Locate chip IC-3 on the scanner's main board, and locate pin 3 on the chip. Clip short one lead of a 470 Kohm resistor and solder it to pin 3. The other end of the resistor connects to position 3 on the counting module. Ground a hook-up wire at any convenient ground point in the scanner, and the other end connects to position 1 (ground) on the module. A lead is run from position 2 and position 4 on the module, and connected to any normally open push button (momentary contact) switch. The push button is used to reset the counter to zero. Install a 'AA' alkaline battery to the module's battery connector, and the job is done. Mounting the module and switch is left to your choice, since the possibilities to suit your needs are endless. The operation is simple, too. Just choose the frequency (channel) you want to count, lock in that frequency, reset the counter to zero, and let it run. Anytime the squelch opens and closes, the count will increase by one. Since the counter will also "see" noise, it may be advisable to set the squelch tight to help eliminate noises other than the calls you wish to count. As always, work carefully to avoid damaging your scanner. If you install a plug on the back of the scanner, and a jack on the module, the module can be used for other purposes.

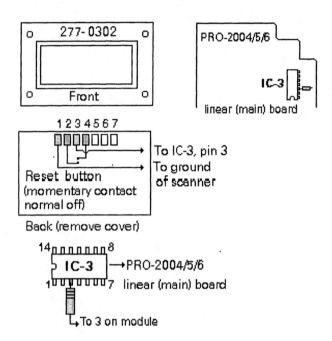

277-0302
Front

PRO-2004/5/6
IC-3
linear (main) board

1 2 3 4 5 6 7
To IC-3, pin 3
To ground
of scanner

Reset button
(momentary contact
normal off)

Back (remove cover)

14   8
IC-3 →PRO-2004/5/6
1   7   linear (main) board

To 3 on module

7

PRO-2022 Mods

PRO-2022 CELLULAR MODIFICATION

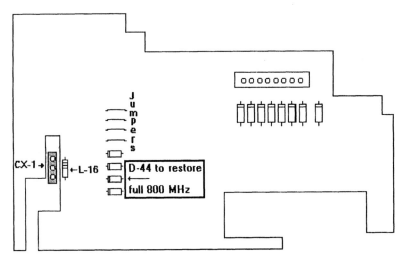

While the **PRO-2022 appears similar** to the PRO-2006, inside they are only distant cousins. The area shown in the drawing is not a discrete board but is part of the main board exposed when the top cover is removed. Remove power and antenna, and looking at the rear of the scanner, remove the two top Phillips-head screws. Slide the top cover towards you, just a bit, until it releases, and, if you choose to, carefully disconnect the speaker. Find the area that resembles the area of the board in the sketch, and remove (or clip) D-44. This is all you need to do to restore Cellular Mobile Telephone (CMT).

NOTE: While I have had no problem with this restoration I have heard of others who were able to enter the CMT frequencies on the display, but were unable to receive the audio.

To restore 66-88 MHz, clip a lead of D-45. There's little to hear in this group, and it may blank 30 to 54 MHz. A little creativity will make this switchable.

PRO-2022 SPEED MODIFICATION

As is, this is a fairly slow-scanning radio but the scan speed can be increased with a little fussy work. You'll need to remove both the top and bottom halves of the case to do this modification, and have at hand a 4 to 5 MHz crystal.

Using the above diagram, locate CX-1 on the circuit board, and from the bottom of the board, desolder all three leads. (This is a good time to use desoldering braid; the solder pads are very vulnerable.) The first one of these I ever did, I wrecked a pad and had to run a jumper from the correct trace! If it happens to you, I suggest 30 gauge wrapping wire for the jumper. There are three leads on CX-1 (ceramic resonator) and only two on the crystal, the center lead being a ground. When you install the crystal, use only the two outside pads. Remember, crystals are shock sensitive so take it easy cutting and bending the crystal's leads. There's lots of room to work on this scanner, so you should have very little trouble (if any!).

PRO-2022 AUDIO MODIFICATION

The PRO-series earphone jacks are attenuated, which is a thoughtful way to protect our ears from melting when you stick those little ear-buds in, but most of us like to control our own level of sound. This mod works with any Pro-series scanner that has the black-white-gray twisted audio harness. At any convenient point on the metal chassis of the scanner, I suggest you install a dedicated ground lug if you intend to do other modifications. It's convenient for working, and a lot easier than locating and making trace grounds. The ground point can be a chassis screw, a lug added through a hole with a nut and bolt (drill carefully!), or just a blob of solder on the chassis frame.

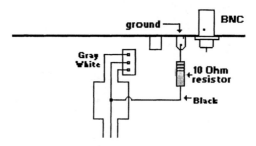

Untwist a couple of turns of the three wire-harness, and from the black wire, just remove enough insulation to solder a 10 ohm resistor between that point and the ground point. With this scanner, you are close enough to both points not to need any hook-up wire, but other scanners may require it.

PRO-2022 LCD CONTROL

The EL (electroluminescent) panels on scanners degrade in a few years (sigh) if they are in constant use. The only apparent way to preserve them is to minimize use. Here's a method to offer some control over the backlight panel in your PRO-2022.

Look at the wiring on the back of the display; find the orange and white wires coming from it. Leave the orange wire alone, but trace the white (negative) wire to a point where you can insert a little circuit. Find a place, probably on the back of the chassis, to mount a center-off single-throw double pole (SPDT) switch and mount it.

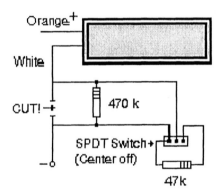

You can build the whole circuit on the back of the switch using 1/4 watt resistors. An option to cutting the white wire is to desolder it where it joins the board, and piggyback the leads there, soldering the white wire to the center pin on the switch and the other wire back to where the white wire was removed from the board.

You now have a choice between *off*, *dim*, or *full light*.

8

PRO-2026 Mods

PRO-2026/2030 CELLULAR MODIFICATION

This modification is so simple someone christened it the "parking lot mod" and that is pretty accurate! Disconnect the power and the antenna, and remove the bottom of the case. With the knobs facing you, in the lower right hand corner you'll find a separate (logic) board. At the top right of the board, there's a bare jumper wire about 1/4 inch long, marked L-201. Snip and slightly separate the cut ends of this wire, for full cellular coverage with 30 kHz spacing.

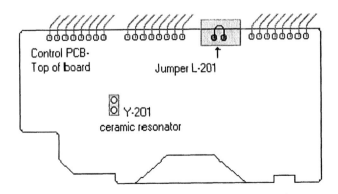

Control PCB-
Top of board   Jumper L-201

Y-201
ceramic resonator

PRO-2026 SPEED MODIFICATION

The options of the 2026 make it an interesting scanner, and the design of the circuit board makes it one of the simpler radios to modify. With a little bit of patience, it's easy to roughly double the scan speed, but be conscious of the fact that there is a proportionate reduction of the delay period as the scan speed is increased. Double the scan speed means half the delay time. See the CMT modification to disassemble the scanner and find the control board. As shown in the drawing, locate the ceramic resonator Y-201. Carefully remove it (it is delicate!) and replace it with a crystal of up to 5 MHz. CAUTION!: too fast a crystal will cause the keyboard to lock up. For these mods, I like to use Radio Shack's 3.57 MHz Colorburst crystal (Part No. 272-1310). When you remove the ceramic resonator, you'll find it has three solder pads. Use only the outer two when soldering in the crystal, and remember crystal leads are sensitive, so trim them with a scissors rather than diagonal cutters. Often that satisfying "snap" signals the demise of the bond between the lead and the crystal.... Occasionally with jobs of this sort, I've installed a crystal socket, just so I could play with the timing.

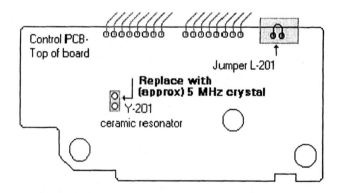

PRO-2026 MEMORY RETENTION #1

If you bought the 2026 to use both as a base and mobile, or for more than one vehicle, chances are that you've lost your memory store. Although I've had users tell me they don't lose memory, for many it has been a problem, so here's an easy fix. At the back of the scanner, above the power input leads, there's a small "mystery hole" in a perfect place to put a small bolt to hold a 9 v battery holder, and that part, plus a 9 v battery snap, are all you need for this mod.

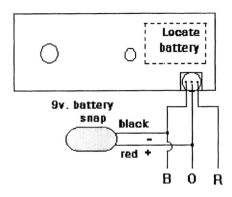

In close to the scanner, strip a small area of the black wire, and do the same to the orange wire. Strip just enough insulation to solder the leads from the battery snap. Solder the black leads together, and the red "snap" lead goes to the orange lead from the scanner. Put a 9 v battery in the battery snap and you no longer need to worry about the scanner dumping the frequencies you so lovingly programmed. From what I've been hearing the battery lasts for a long, long time. When you change it, do it with power connected, and there'll be no chance of loss.

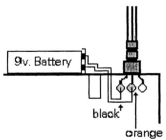

PRO-2026 MEMORY RETENTION #2

Another method of wiring the PRO-2026 for memory might be neater, and offers a further option, as well. This is how I do the mods that come across my bench, and it's my preferred method. Remove the top cover and see where the power cord enters. See where the three wires are soldered to the board. This is where the work is done.

Above and a little to the left is a small hole, just big enough to lead the wires of a 9 v battery snap through. Right at board level, remove a little insulation from the end of both the black and orange wires, and solder the red lead from the battery snap to the orange lead in the scanner. Solder the black lead from the snap to the black lead at the circuit board and the job is done. I secure the battery to the scanner's back with a bit of double-face tape. The further option I mentioned is to locate the battery and all inside the scanner (there's plenty of room!) but if you do it make sure the battery is well insulated and secured.

9

PRO-34/37 Mods

PRO-34 CELLULAR MODIFICATION

If I was humble, then in my humble opinion, this is the first of the modern handheld scanners. It has everything the ordinary scanner user needs, without the frills, and is very well made, and the internal shielding is excellent. Like most handhelds, it takes a bit of work to"dissect" it to the point of our modification. Pull the knobs, antenna, and battery pack. Remove the four small Phillips-head screws on the back. Pulling up from the bottom of the scanner, start to separate the front and case halves. You'll find that the back of the case needs to be eased over the knob/antenna assembly. Now it's time to do a bunch of desoldering. I have an old "solder sucker" that I use for this part of the job, and it works well, but if you're used to braid use that. There's a lot of solder! (SEE INSERT)

Now remove the board connectors, the four brass hex screws, and gently lift the board straight up (there's a connector underneath). You'll find a metal RF shield underneath held by three small screws, and possibly a ground wire (varies), and under this is the logic board. As shown in the drawing, find the diode matrix and remove

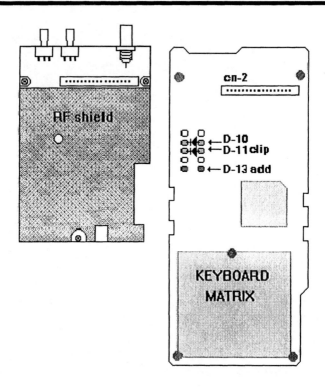

(or snip) D-11. Check to see if pad D-13 has a diode. If not, solder in a 1N914. With care (and tinned leads) you can solder this diode in from the top of the board. Carefully reassemble everything, and program for cellular in a dedicated search band. The PRO-34, like all PRO- models, defaults to the necessary 30 kHz spacing.

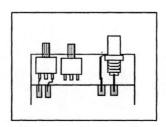

PRO-34 SQUELCH MODIFICATION

Most, if not all, PRO- series scanners have a lot of slop in the squelch control. I don't find it a problem, but since the question has been asked, it's fairly easy to "tighten" it up if you want, like this...

Remove the back, doing the usual routine, and on the linear pc board locate chip C-101. Count around to pin 12. Simply remove the resistor between pins 12 and 14 (R162- 27 Kohm), and replace it with a resistor with a value of 150 to 200 K.

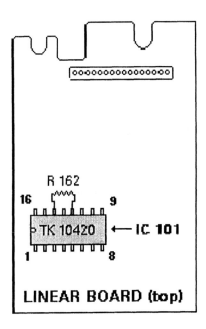

Remember to be able to undo your work. Sometimes a mod just isn't what you want, or very possibly, you might want to change a value to effect a greater or lesser result.

PRO-34 SPEED INCREASE

Although it calls for working in close quarters, it's fairly easy to effect an 80% increase in scan speed on this model. At the right side of the logic board, roughly midway between the CN-2 connector and the CPU, you'll find the ceramic resonator (CX-1) that determines the scan rate. It has to be removed and replaced with a crystal. Remove the battery, antenna, and knobs, and go into the logic board, the same as for the CMT mod. At the right side of the board, halfway between the CN-2 connector and the CPU, locate CX-1 and remove it. You'll find it has three leads on the pads, but you can disregard the center one. Install a crystal of 3.5 to 5 MHz in its place. Trim the crystal's leads short enough to let you make a right angle bend into the two outside pads at CX-1's former location. If you tin the leads on the crystal and work gently, you can easily solder the new crystal in from the top of the board. Insulate the crystal housing to prevent shorts.

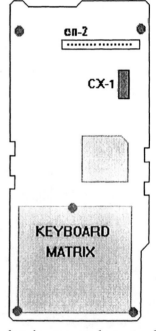

There is a price for this mod. As the scan speed is increased, the delay time is shortened proportionately. The greater scan speed is a definite plus in areas where trunk systems are used, but increasing the speed too much will cause the CPU to lock-up, and "freeze" your keyboard.

PRO-37 CELLULAR MODIFICATION

This is a well made scanner! A generation up from the PRO-34, it is very similar physically, but there are some circuit changes. To disassemble, refer to the PRO-34 CMT mod. To restore the "missing" part of the 800 MHz spectrum, it's only necessary to locate D-13 and either clip or remove it. Of course, you must not listen to the 869.00 to 892.00 MHz part of the band! Work carefully, and watch the soldering tool, plastic melts in a hurry!

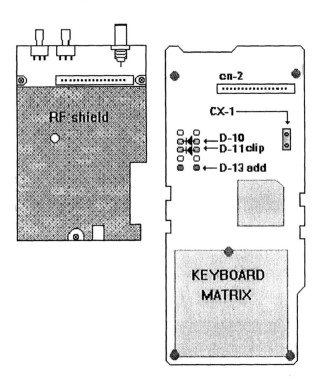

PRO-37 SPEED MODIFICATION

Although it calls for working in close quarters, it's fairly easy to effect an 80% increase in scan speed on this model. At the right side of the logic board, roughly midway between the CN-2 connector and the CPU, you'll find the ceramic resonator (CX-1) that determines the scan rate. It has to be removed and replaced with a crystal. Remove the battery, antenna, and knobs, and go into the logic board, the same as for the CMT mod. At the right side of the board, halfway between the CN-2 connector and the CPU, locate CX-1 and remove it. You'll find it has three leads on the pads, but you can disregard the center one. Install a crystal of 3.5 to 5 MHz in place of CX-1. Trim the crystal's leads short enough to let you make a right angle bend into the two outside pads at CX-1's former location. If you tin the leads on the crystal and work gently, you can easily solder the new crystal in from the top of the board. Insulate the crystal housing to prevent shorts.

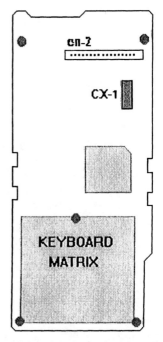

There is a price for this mod. As the scan speed is increased, the delay time is shortened proportionately. The greater scan speed is a definite plus in areas where trunk systems are used, but increasing the speed too much will cause the CPU to lock-up, and "freeze" your keyboard.

10

PRO-39 Mods

PRO-39 CELLULAR MODIFICATION

The PRO-39 is built on much the same frame as the PRO-43 but has major circuitry differences. Disassemble the same way as the PRO-43, and at the left center of the logic board, find the diode matrix. Clip or remove D-6 to restore cellular mobile frequencies. See next page for a diagram.

Like all the PRO- series, when this scanner is opened up to the "missing " frequency, the CPU will default automatically to the 30 K spacing needed to match the CMT usage. To test, after the scanner is reassembled, program a dedicated search bank with 869-894 MHz. Be aware that it is a violation of the Electronic Communication Privacy Act (ECPA, 1986) to listen to cellular conversations!

PRO-39 SPEED MODIFICATION

It's possible to speed up the scan/search rate in many scanners, and the PRO-39 is no exception. Remember, there is a cost for doing this, a proportional loss of delay time. Modern scanners run pretty fast as is, but you might see an advantage if you do a lot of trunk monitoring.

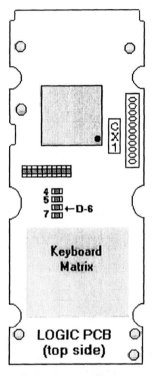

To open this handheld, see the 800 CMT mod. Look at the drawing to locate ceramic resonator CX-1 on the logic board. It has to be removed (close work!) and rather than pulling up the logic board to get to the back side you might want to simply crush it (gently), leaving the three pins under it in place. Replace it with a crystal with a value no higher than 10 or 12 MHz. More will probably cause the CPU to lock up. Connect the crystal to the two outer pins. NOTE: Be very careful snipping and bending leads on crystals, and most other components. The shock of that satisfying "snip" can harm sensitive "innards."

The center pin under a ceramic resonator is a ground, and it is a good idea to ground the case of the crystal to the pin using a little 30 gauge wrapping wire. I've done the job both with and without the ground and never had a problem either way, but a little insurance never hurts.

Here's that reminder again....The author and publisher take no responsibility for any damage done to equipment or person as a result of these modifications. Any modification to any equipment can void your warranty.

PRO-39 AUDIO FIX

Like the PRO-43, the sound of this scanner pleases some, and bothers others. To induce a little treble and sharpen the sound, here's a couple of fixes to try. See the 800 CMT mod for directions to open the scanner for work. Go to the middle (AF) board and find IC-204, in the corner of the board. Flip the board over to the bottom side, and where the pins of IC-204 penetrate, locate the cluster of components. See the drawing to locate C-235 and remove it. This is a 0.015 µF capacitor. If you choose, you can replace the surface mount cap with a regular 0.020 µF cap. I've done this on a couple of scanners, and usually my customers tell me that it sounds better. My old, worn out ears can't tell the difference, but yours might.

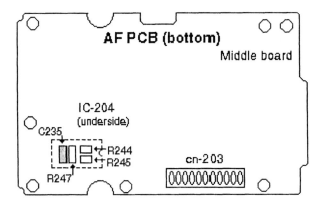

11

PRO-43 Mods

PRO-43 CELLULAR MODIFICATION

This is the hottest handheld it has been my experience to work with, and here's how to replace the "missing" portion of the 800 MHz band. Remove the battery pack, the antenna, and the knobs. Take the four screws out of the back, and from the bottom, gently ease the case open. Remove the six small silver screws holding the "top" board. Desolder the antenna ground pad. Just "south" of it, the antenna lead comes through from the lower board, and it too must be unsoldered. You can choose whether or not you want to disconnect the two wiring harness plugs; there is enough lead to leave them connected.

Lift the top board away from the second, paying attention to the black pin connector near the bottom of the assembly. Remove the two silver screws from the second board, and lift it away from the third (logic) board. Pay special attention to the two white header connectors at the side. This board also contains the controls. Desolder the exposed metal RF shield at all four points, and remove it. Notice its orientation.

Find the row of diodes as shown in the drawing (they are surface mount chips, and as such are very tiny). Remove D-4. (Carefully, see the note.) Personally, I like to use a 25 watt soldering pencil for this; I think more heat used quickly transfers less heat to the board. Save D-4 and store it, should you ever want to undo the mod.

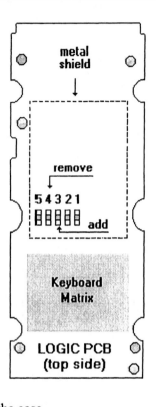

Note: This diode can be installed at D-3 to restore 68 to 88 MHz, mostly unused in this country except for some out-of-channel TV audio. Diodes are polarized so remember to keep the same orientation. In a pinch, the diodes can be replaced with 1N914's. Don't mess with D-5, it is the 30 KHz default switch. As you reassemble the scanner, don't forget the connectors, and be careful with the wire group that comes up the right side of the chassis. Fit it carefully or you will have trouble closing the case.

Diode allocations: D-1 unknown, D-2 enables 30 to 50 MHz (when present), D-3 enables 50 to 90 MHz (when present), D-4 enables 825 to 870 MHz if missing, D-5 defaults to 30 kHz.

PRO-43 SPEED MODIFICATION

If you are in an area with trunked systems, this modification might suit your needs very well. With the fast switching of channels by the site control computer, a faster scanner equals more calls.

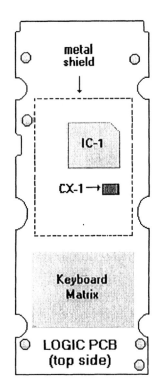

There is a price, though. As the scan speed is increased, the delay time decreases in proportion. The beep you heard will turn into a chirp, and delay time will drop to less than a second. The component CX-1 is a ceramic resonator that controls the scan rate. In the PRO-43, it runs at 4.19 MHz. With a little "fittin 'n fiddlin" it can be replaced with a 10 MHz crystal, to give a speed increase of more than 100%.

Use the disassembly method shown in the Pro-43 CMT mod, and find CX-1 on the logic board. Very carefully, remove it. You'll find it has three pads, but in installing the crystal, only the two outside ones are used. **Caution:** the CX-1 is delicate. Also, when you clip and bend the leads on the crystal to fit, be very gentle, the crystal's leads are sensitive to shock. Don't forget a plan to undo your work!

PRO-43 AUDIO MODIFICATION

The PRO-43 is a good scanner, but the audio has a tendency towards a "tinny" sound. This is a modification to improve the sound. On the foil (back) side of the middle (AF pc board), using the drawing find C-341. It's a small surface-mount capacitor, and is unmarked. On the top side of the same board is IC-304. The cap is directly under it. The cap can simply be removed for some improvement. To really make a change in clarity, in place of the now absent chip capacitor, replace it with a .002 µF disk capacitor. By clipping the leads short and using the small caps that Radio Shack packages, you can make an easy fit. The cap to use is marked "202." Other caps can be used in this position to alter the audio to your taste.

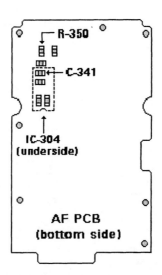

Note: When I need to try a variety of components under power in a surface-mount context, I solder 30-gauge wrapping wire to the pads in question and bring the leads outside the assembled scanner to do my testing. Radio Shack's part number is 278-501/2/3 depending on your preference for color.

PRO-43 REDUCE BEEP

This mod is a happy medium between "To beep or not to beep!" for your PRO-43. Use the disassembly text from the PRO-43's cellular mod, and get to the second board (AF pcb). On the top side, as shown, locate R-363, a surface-mount resistor, marked 471 (470 ohms). Carefully remove it and the "beep" will be greatly reduced.

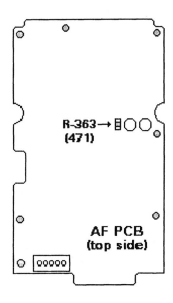

PRO-43 BEEP DELETE

I've gotten letters asking how to delete the beep on the PRO-43 and here's how! I like the confirming sound but the choice is yours. A little ingenuity will allow you to make this switchable. Using the disassembly method for the CMT mod, work your way to the logic board, and find CN-3 (10 pin connector) and separate the parts. Bend or otherwise disable pin 6 and the beep is a thing of the past.

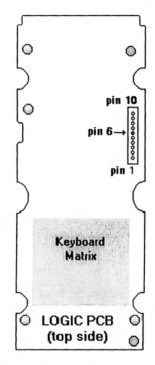

PRO-43 S-METER MODIFICATION

Find a spot for a miniature plug in your PRO-43 and you can install an "S" meter for signal comparison. Look at PRO-43's CMT mod for disassembly, and work on the back (solder side) of the AF board. As in the drawing, just bring out a lead from the cathode side of D-302, and a ground, connected at any convenient point. Connect these leads to the mini-plug.

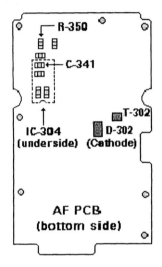

You'll need to locate the mini-plug inside the PRO-43, making sure there's enough clearance to close the case. The meter and meter circuit can be built inside a stand-alone box and, with an appropriate jack, the meter can be used on other scanners. If you're good at really close work, you can fit the meter circuit components inside the 43, but it is much simpler to use an external box.

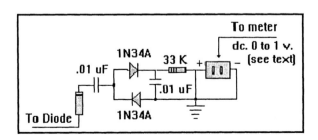

12

PRO-46 Mods

PRO-46 CELLULAR MODIFICATION

There are some differences in the building of this radio, as well as a different approach to blocking those missing frequencies.

Disassembly: Remove the antenna and battery cover, and the four black screws from the back. The back is "latched" at the bottom and must be eased up from that point to open. The case screws are all that hold the PC boards in place. Gently separate the first board from the connector at the right edge of the board (just below the center line) by prying up with a thin blade. Avoid force. The RF shield on this scanner is just a copper flashed plastic membrane, and care needs to be taken when you desolder it. Simply stay on the copper surface where the solder joints are, and keep the heat spread to a minimum.

The mod: At the right side of the board, locate the two surface mount chips (marked "0"). These are jumpers, and must be removed with care to avoid damaging them. You'll need at least one good one for the mod. Slightly to the right of this area, you'll see two sets of blank pads. Use one of the saved jumpers to install at the top pair of pads to restore the missing 800 MHz frequencies, and change mode when appropriate to 30 K spacing. The other (lower) pair will enable 68 to 88 MHz, but they will disable 30 to 54 MHz. It would be possible to make this jumper switchable, but there's little point. Test by dedicating a search channel, using low 869.00, high 894.00. The scanner will default to the correct frequencies. Remember, listening to CMT is illegal, although modifying the scanner is not...

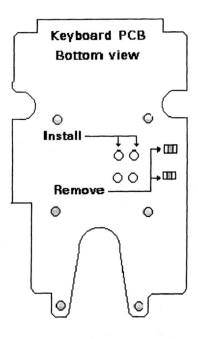

If you have a disaster and lose the jumper chips, the pads can be jumped with a small piece of "wrapping" wire. Watch where you point your "iron," plastic melts quickly!

PRO-46 SPEED INCREASE

As trunking systems are being used more, a faster scan speed can be advantageous, and in this scanner it is not hard to nearly double the manufacturer's built-in rate. I don't have the value for the ceramic resonator being used, but judging from its slow speed a replacement crystal in the 5 MHz range should do the job nicely. Too high a value will cause the keyboard to lock up, so err on the conservative side, possibly with a 3.57 Meg crystal to be safe. The disassembly for this scanner is on the cellular mod page. Locate the resonator as shown in the drawing. It's Y-201, and is marked 800J. Carefully remove and save it (it is delicate, so go softly). Replace it with a crystal, cutting and fitting the leads to match the pad. Remember, crystals are sensitive to shock, so cut and bend as gently as you can.

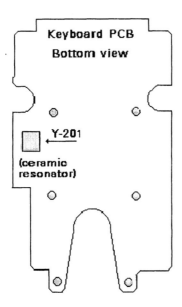

Tin the leads on the crystal and you'll be able to solder in the new crystal from the top of the board. Should you care to do it, a crystal socket can be installed, giving the option of interchangeable scan speeds. Crystals can be readily ordered from Radio Shack.

NOTE: As scan speed increases, delay time decreases in proportion.

PRO-46 BEEP MODIFICATION

Personally, I like the confirmation that the beep awards, but others tell me they find it objectionable, so here's how to disable it. Disassemble the scanner using the method shown in the CMT modification. Locate the CPU, IC-1489, and find pin 80, the first pin on the left counting up from the corner. Follow the trace from there to L-204. With a very sharp blade (I use a scalpel, but any of the commercial razor knives will do the job), cut the trace anywhere convenient. This will delete the beep. If you need to restore it, it's simple to scrape a little of the covering off the trace and solder bridge the cut. With a little creativity, this modification could be made switchable.

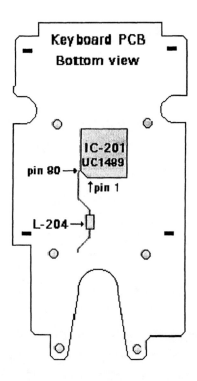

13

BC-100XLT
BC-200XLT
BC-205XLT Mods

BC-100/200XLT LIGHT MODIFICATION

There are many scanner users of BC's who would like to have the display light stay on, so here's how to do it. Disassemble the scanner as shown in the BC CMT modification. Near the bottom left side of the CPU (UC 1147), find transistor Q-201 and carefully remove it. For a light that's always on (it will reduce your battery life by about three hours per charge) just solder a bridge from pad "A" to pad "C".

A little more complicated, but better option, is to add a switch by running leads from pads A and C to a mini switch mounted in the space near the squelch control, giving the choice of always on or auto-switch. Changing the resistor R-262 to a value between 220 to 470 ohms will give a dimmer light, but will extend battery life.

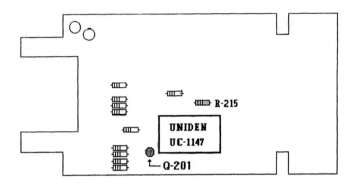

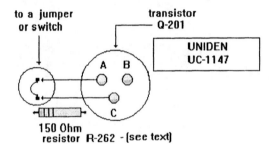

BC-100/200XLT EXTENDED BATTERY LIFE

You can get a few more hours per battery charge with this mod, as long as you are willing to give up the low battery warning. With this mod the scanner will just run out of juice. Disassemble the scanner as shown in the CMT modification. As per the drawing, remove and save D-203 and R-208. Reassemble.

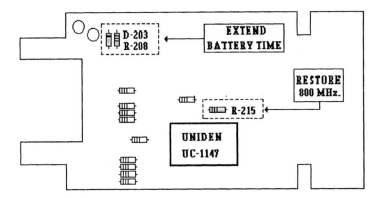

BEARCAT EARPHONE MODIFICATION

The Bearcat series, like the PRO-series, have the earphone jacks attenuated, to protect our hearing I suppose, or maybe to avoid blowing out those little earphone buds. At any rate, here's a way to bring up the audio to full level. Use caution until you know the level of volume you have in the phones; the difference is significant! This mod works on the BC 100/200 and will likely work on other BC handhelds.

Take off the power pack and antenna, and remove the screws holding the back. Remove the screws holding the contact spring. Open the back. If you look at the earphone jack, on the left side you'll see access to the inner ground contact, and at that point you need to solder a short piece of jumper wire. The other end of the jumper goes to the grounding lug of the BNC antenna connector. Put things back together and enjoy the increased volume.

BC-200XLT CELLULAR MODIFICATION

There's a little confusion regarding this mod, but as far as I know, this is the one correct modification for this scanner.

Disassembly: Remove the battery pack and antenna, and the two back screws. Remove the two screws that hold the battery retainer spring. Remove the spring. Gently ease open the rear case, and remove the two small screws that hold the circuit board. Separate the front panel from the mainframe. Locate the 64 pin CPU, marked "UNIDEN- UC 1147." Just above the "DEN," there's a 10 ohm resistor (R-215) surface chip that needs to be removed. It can be cut in place with wire cutters, but my own preference is to carefully unsolder it, and store it. (I have found other uses for these and other chips, and usually stick them to a piece of cellophane tape, and park them inside the scanner.) Removing the resistor completes the mod.

Reassembly: Carefully fit the top of the front face into the groove just under the controls, and line up the connector at the bottom of the board with the mating socket and press them together. Line up everything, and replace the screws.

NOTE: Bearcat scanners do not default to 30 kHz spacing when in the CMT part of the spectrum, so many calls will be skipped.

BC-205XLT CELLULAR MODIFICATION

There is a difference between restoring the BC-200 and the BC-205, even though they appear identical. Here's the method for the 205. Disassemble as shown in the 200 XLT CMT mod. Locate R-215. Do not remove or cut. Instead, add a 10 Kohm resistor between the left end of R-215 and pin 8 of K1013, close above it. This is all that's needed to restore CMT in this scanner. Carefully reassemble the scanner.

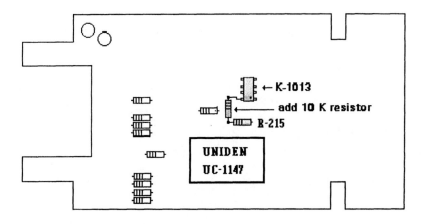

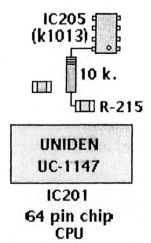

BC-200/205XLT SPEED MODIFICATION

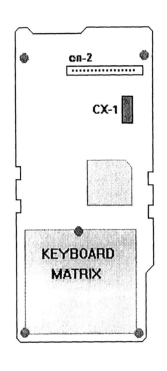

There's a way to more or less double the scan-search speed in your BC, but the price as usual is a shortened delay time. There's another problem, too. Y-201 (ceramic resonator) runs at 400 kHz, unlike the MHz crystals used in other scanners, and a higher value crystal or resonator can be hard to find. One possible source is Murata-Erie in Smyrna, GA. Call 800-368-8370 for availability. Bill Cheek, author of the *Scanner Modification Handbooks* and *The Ultimate Scanner* has a supply of 800 kHz resonators for this scanner. E-mail: bcheek@san.rr.com

Open the scanner as described in the cellular mod, and remove the two screws at the bottom of the visible circuit board. Carefully pull the board, noting the connectors, and expose the logic board. Find the little cube marked Y-201 (TT665), and replace it with an 800 kHz crystal or resonator.

14

Other Bearcat Mods

BC-140/142 MEMORY MODIFICATION

This appears to be one of those transitional scanners, where Uniden used a future CPU in an early model, and blocked out the model coming up. It is not possible to add 800 MHz to this model, but you can add channels. This is how it's done. From the bottom, remove the four case screws, and separate the case  halves. Proceed gently, there are wires connecting both parts.

At the corner of the board where the connectors are, look for the "UNIDEN" 30 pin chip. Near this chip find a diode with insulating tubing on its leads. The leads go from the top of the board to the underside. Clip or remove this diode to enable 16 channels. Carefully re-assemble the scanner.

BC-700A CELLULAR MODIFICATION

This scanner is a hybrid of some other Bearcats, I think, but it can be "gifted" with the CMT frequencies by doing a little close work. When it's done, it won't default to 30 MHz, but that's typical of Bearcats. It will see any calls that turn up at 12.5 KHz spacing. Remove the power, of course, and the antenna, and then take out the case screws. Open the case and disconnect the speaker plug (Carefully!). Turn the scanner on its back, and with the face away from you, take out the screws holding the faceplate. Tilt the faceplate enough to unplug the connector on the clear cable harness so you can tip the faceplate all the way. There is a plastic, copper flashed membrane RF shield that needs to be removed by unsoldering. Be careful where you put the heat! As shown, remove or clip the chip resistor marked 472. Carefully resolder the RF shield and reassemble the scanner. (don't forget the connectors.) Test by entering into search 869.000 low, 892.000 high.

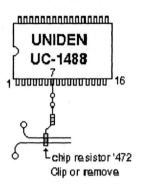

The ECPA prohibits listening to these frequencies, but will allow the modifications if done by the scanner owner... [*Publisher's note: well, we think so, but see page 165.*]

BC-760XLT CELLULAR MODIFICATION

This model requires a bit more work than most to restore CMT, but when you're done, and it works, it's a great feeling! At first glance, it looks complex, but if you go a step at a time, it isn't nearly as hard as it looks. There are two different possibilities with the circuit board, and one couldn't be simpler. Look to see if the connector for your antenna is a BNC. If not, all that's required is that you clip pin 20 on the CPU. If it uses a BNC, get ready for a little close work. Disconnect all power, and remove the antenna, and then turn the scanner bottom up. Remove the four screws holding the bottom cover, and set the cover aside. Turn the scanner so that the front panel faces you, and find the SANYO IC. Find the row of pins above the IC, and from the left end count to pin 26. With a razor knife, cut the two traces that lead to pin 26. With solder, bridge pins 19 and 20. At the same time, bridge pins 26 and 27. (Careful where you put your soldering tool, plastic melts fast!) Solder a bridge across the two leads of the chip transistor just above pin 27. Locate the chip resistor marked 472 and remove or cut it. Reassemble the scanner, and test by entering 869.000 low, 892.000 high in search. Bearcat models do not default to 30 kHz spacing, so some calls will be skipped.

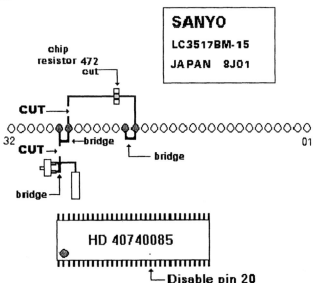

BC-855XLT CELLULAR AND 100 CH MOD

A few scanners seem to be transitional stages of future models, and this is one of them. With very little work, the capability of the 855 can be doubled, and will store 5 banks of 20 channels each, along with returning the CMT portion of the spectrum. (Remember, you can have the capability in your scanner, you just can't legally listen to it!) From the bottom of the scanner: remove the 5 screws, and separate the case halves. Unplug the speaker, and lay the case halves side by side. Find the jumpers as indicated in the drawing and clip E and D. (Either or both can be cut to get desired results.) As shown in the drawing, add a 10 K resistor to the circuit. Both jumpers can be soldered to the same resistor. Reassemble and test by searching 869.000 MHz to 892.000 MHz. **NOTE:** The Bearcat scanners do not default to 30 kHz, the spacing allotted to cellular phones, so some calls will be skipped.

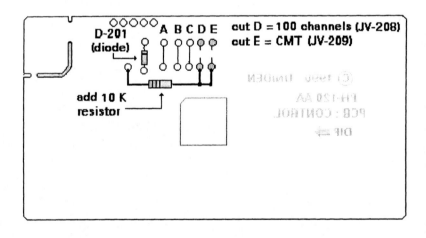

BC-890XLT 800 CELLULAR MODIFICATION

This is the only Bearcat scanner I know of that will default to 30 kHz when the CMT modification is done. 30 kHz spacing is required in the CMT mode because that is the spacing determined between cellular frequencies. Other spacing results in most calls being skipped over. Remove power, and take the screws out of the cabinet (there are ten), and separate the cabinet halves. Unplug the speaker connector. There is one bracket screw at the middle of the main board, and four screws holding the faceplate. Remove these, and manipulate the faceplate shield so you can tip the faceplate, and uncover the logic board. Disconnect the ribbon connectors, and connectors J-4 and J-5. Remove the logic board with the faceplate. As shown in the drawing, find the CPU marked UC-1514. Just to the lower right of the CPU, locate the three chip resistors, and carefully unsolder the closest one to the CPU (marked 104). This chip needs to be removed without damage, so you can relocate it at the solder pads indicated, right by the memory battery. Reassemble the scanner, remembering to reconnect all the wires (including the speaker!). Test by programming into search: 869.000 low, 892.000 high. Note: it is legal to modify your scanner; however listening to it is a violation of the Electronic Communications Privacy Act.

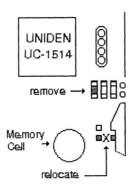

BC-950XLT CELLULAR MODIFICATION

This model requires a bit more work than most to restore CMT, but when you're done, and it works, it's a great feeling! At first glance, it looks complex, but if you go a step at a time, it isn't nearly as hard as it looks. There are two different possibilities with the circuit board, and one couldn't be simpler. Look to see if the connector for your antenna is a BNC. If not, all that's required is that you clip pin 20 on the CPU. If it uses a BNC, get ready for a little close work. Disconnect all power, and remove the antenna, and then turn the scanner bottom up. Remove the four screws holding the bottom cover, and set the cover aside. Turn the scanner so that the front panel faces you, and find the SANYO IC. Find the row of pins above the IC, and from the left end count to pin 26. With a razor knife, cut the two traces that lead to pin 26. With solder, bridge pins 19 and 20. At the same time, bridge pins 26 and 27. (Careful where you put your soldering tool, plastic melts fast!) Solder a bridge across the two leads of the chip transistor just above pin 27. Locate the chip resistor marked 472 and remove or cut it. Reassemble the scanner, and test by entering 869.000 low, 892.000 high in search. Bearcats don't default to 30 kHz spacing, so some calls will be skipped.

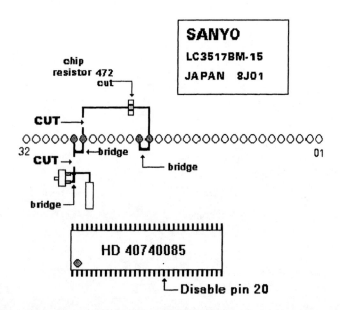

BC-2500XLT CELLULAR MODIFICATION

Remember, it's illegal to listen to this locked-out band, but it is OK to modify your own scanner (*we think, see page 165*). This is, as far as I know, the only scanner in the BC-XLT series that will default to the correct 30 kHz spacing to properly receive CMT calls.

As usual, remove the battery pack and antenna. Lay the scanner face down. Take out the two top black Phillips-head screws, and the two plated screws at the bottom. Gently pry the case open. Carefully lift the exposed circuit board, and separate it from the white (Molex 11-pin) connector. Don't pull on the ribbon connector.

Find the CPU chip, UC-1513A. Under it you'll see a group of surface-mount resistor chips, marked 104. As in the drawing, carefully remove the chip at the lowest left side position, and move it to the exposed pad at the left. Solder in place. Re-assemble the scanner, being careful not to bend any pins on the Molex, and test by programming 869.00 MHz low, 892.00 high, into the search function.

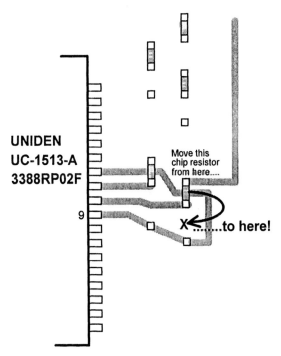

UNIDEN
UC-1513-A
3388RP02F

9

Move this
chip resistor
from here....

X ....to here!

RESET AND TEST PROGRAM

Most scanners made by Uniden, including those with the Radio Shack label, have a **special memory-clear/reset** procedure whereby the scanner's memory is totally cleared and the first 20-25 channels are filled with special test frequencies. For the most part, these frequencies are nothing special, but there can be instances where the procedure is worth knowing.

There is no harm done by this procedure other than it will erase the current contents of memory and replace it with the test frequencies, so don't do it unless you're all set for an extended programming session thereafter.

Reset for Bearcat scanners:

Press and hold 2 9 manual together,
Turn scanner on.
Turn scanner off.
Press and hold 2 9 scan together,
Turn scanner on.
This action will load channels 1 through 25 with test frequencies,
and clear the remaining channels.

15

Regency Mods

R-4030 CELLULAR MODIFICATION

There's a little confusion regarding this mod, but as far as I know, this is the one correct modification for this scanner.

Disassembly: Remove the battery pack and antenna, and the two back screws. Remove the two screws that hold the battery retainer spring. Remove the spring. Gently ease open the rear case, and remove the two small screws that hold the circuit board. Separate the front panel from the mainframe. Locate the 64-pin CPU, marked "UNIDEN- UC 1147." Just above the "DEN," there's a 10 ohm resistor (R-215) surface chip that needs to be removed. It can be cut in place with wire cutters, but my preference is to carefully unsolder it, and store it. (I have found other uses for these and other chips, and usually stick them to a piece of cellophane tape, and park them inside the scanner.) Removing the resistor completes the mod.

Reassembly: Carefully fit the top of the front face into the groove just under the controls, and line up the connector at the bottom of the board with the mating socket and press them together.

Line up everything, and replace the screws.

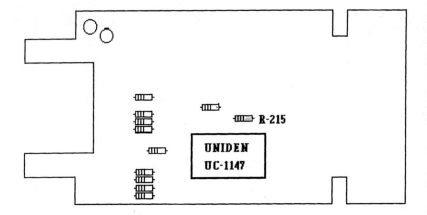

NOTE: Scanners in the Bearcat series default to 12.5 kHz spacing in the CMT part of the spectrum, so many calls will be skipped.

R-1600 CELLULAR MODIFICATION

This model requires a bit more work than most to restore CMT, but when you're done, and it works, it's a great feeling! At first glance, it looks complex, but if you go a step at a time, it isn't nearly as hard as it looks. There are two different possibilities with the circuit board, and one couldn't be simpler. Look to see if the connector for your antenna is a BNC. If not, all that's required is that you clip pin 20 on the CPU.

If it uses a BNC, get ready for a little close work. Disconnect all power, and remove the antenna, and then turn the scanner bottom up. Remove the four screws holding the bottom cover, and set the cover aside. Turn the scanner so that the front panel faces you, and find the SANYO IC.

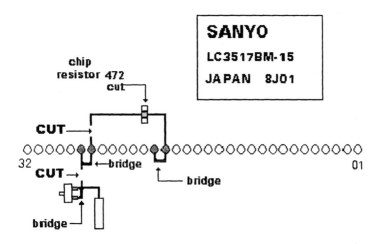

Find the row of pins above the IC, and from the left end count to pin 26. With a razor knife, cut the two traces that lead to pin 26. With solder, bridge pins 19 and 20. At the same time, bridge pins 26 and 27. (Careful where you put your soldering tool, plastic melts fast!) Solder a bridge across the two leads of the chip transistor just above pin 27. Locate the chip resistor marked 472; remove or cut it.

Reassemble the scanner, and test by entering 869.000 low, 892.000 high in search. Bearcat models do not default to 30 kHz spacing, so some calls will be skipped.

16

Mod Projects for All Scanners

CARRIER OPERATED RELAY

If you are an owner of a PRO-2004/5/6, this is an easy switch to hook up. For other scanners, see the next page. The circuit is really simple. (Suits me!) The squelch on/off automatically makes everything go by taking its signal from the control chip on the main board. You can power this from a 12 v tap inside the scanner, or you may opt to use a separate power supply, or batteries.

I used 2N2222 NPN transistors with good results, but just about any common NPN should work. All the parts except the relay are available from Radio Shack. The relay has to be pretty sensitive, a low current model with a 20 Kohm coil resistance. Build this circuit on a little bit of perf board, which need not be much larger than the relay's base. Nothing is critical except the possibility of overheating the transistors. To avoid that, you might want to use a couple of sockets.

There's a nice feature imbedded in the use of this switch. Since it only operates on an active carrier, you will be able to monitor either a bunch of frequencies, say 800 MHz trunk, or a single channel to see frequency of operation. Any recorder with a remote switch can be used, or, since the remote portion of the switch is isolated after the relay, other devices can be operated as long as the load doesn't exceed the relay's contact rating.

See Appendix 3, **Advanced Data**, for the squelch/scan control pins in other scanners.

CARRIER ON/OFF INDICATOR

This simple circuit can readily be added to your scanner to indicate scanner status by showing a green LED as long as there is no carrier apparent to your scanner. When your scanner "hears" a station, the circuit turns off the green LED and turns the red LED on. Along with helping to satisfy my lust for flashing lights (a weakness!), it works as an indicator to let you know if the radio is locked on a silent carrier, and acts as a reminder to let you know if the scanner is on, even though the volume is turned to zero, as happens sometimes in my home. (Someday I need to build a phone switch!)

I built this circuit on a tiny piece of perf board, and mount it in the same case as one of my meters, which in turn is mounted on top of my PRO-2006. Actually, the LEDs are built into the meter face. Since the meter case is mounted directly on my scanner, I brought the leads for the circuit board through a hole in the case. You can suit your own needs. The circuit is so simple, it can be built inside the scanner without a board. The needed parts are readily available:

```
transistor    Q-1  NPN, 2N2222A
resistors     R-1  1k              R-3  3.3-k
              R-2  470             R-4  33-k
```
(all resistors are common 1/4 watt)

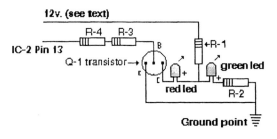

Add a small perf board, and 1 red, 1 green LED (T-1 size, drill 1/8 inch hole). The signal lead hook up point varies: PRO-2004/5/6/2021, connect to pin 13 of IC-2. PRO-2022, pin 13 of IC-1. Bearcat 200/205, anode of D-7. Bearcat 760/950/855/890, anode of D-15.

The 12 volt power lead: PRO-2004, connect to R-229 (either side). PRO-2005/6, PRO-2022, connect to R-229 (either side). PRO-2021, R-143 (either side). Bearcat and Regency scanners: connect to the low side of the power switch. Ground lead goes to any ground.

BFO FOR YOUR SCANNER

If you want to listen to Morse code, this little circuit may be just what you need. The beat frequency oscillator (BFO) injects a signal at or near the intermediate frequency of your scanner (455 KHz) and causes the incoming code to heterodyne, so you hear a clear sharp signal, rather than the "zupp-zupp" sound you hear without it.

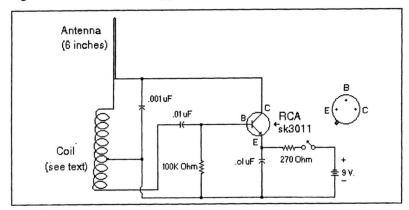

The only faintly exotic part needed is the tuning coil, which is 40-300 µF. You may need to prowl the electronic stores to find one. Try to get J.W. Miller type 2002 or its equivalent. The whole circuit can be built on a piece of perf board, and there's nothing critical in the wiring.

The antenna can just be a short length of hook-up wire, or you can use an antenna saved from a junked transistor radio. It only needs to be about 6" long, and you can play with the length until you get the best results. When everything is assembled, adjust the slug in the antenna coil to the center of the winding, and lock the scanner to a strong station. Adjust the tuning slug until you hear a "whistle" along with the station you are listening to.

Now retune the scanner to other stations. When everything's right you should hear the whistle with every frequency. This may take some "fiddlin"! You are trying to tune the BFO to exactly 455 kHz, and the slug will cover frequencies from 300 kHz or so, right up to about 1500 kHz. Hint: If you hear the tone behind just one station, you are tuned to a higher frequency than you want. Once you get the right setting, leave the tuning slug in place, and you can use the BFO with just about any radio. For example, you can use it with your shortwave receiver. If you find the BFO blocks weak signals, just shorten the BFO's antenna, or move it further from the scanner.

The BFO can also be used to find weak stations if you have a pretty good idea of the frequency you are looking for. Using the search feature on your scanner, bracket the station's frequency and turn on the BFO. Listen for the BFO tone to be on both sides of a frequency and look in the center for the actual station. Without an adequate antenna you might not hear the station, but you'll at least know it's there. Another tip: An easy way to be sure you are wiring the tuning coil correctly. With an ohmmeter, look for lower resistance between the leads of the shorter coil.

MATRIX SWITCHING: PRO-2006 & OTHERS

A persistent problem is silent carriers that lock the scanner, and prevent the jump to the next channel. The problem isn't severe as long as you're next to the set, but if you are at any distance, resetting the scan can be a nuisance. This is a lazy way to solve the problem.

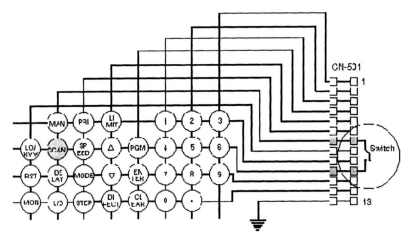

The drawing shows the switching operation of the keyboard. Every function is represented by a combination of connections that run through connector CN-501, and you can quickly trace the 2-pin connection that "turns on" that particular function. The switching is done by momentary contact, so any device that will make that connection can be "piggy-backed" to those leads to perform the same effect as pushing the keyboard button.

By using infra-red or radio control, any type of switching from the keyboard can be controlled remotely. All you need to do is find the pins that suit your needs and bring them to a jack on the exterior of the scanner and add your switching device's receiver.

I use this gimmick with R/T's salvaged from broken radio control cars, electronic doorbell devices, and a primitive infra-red R/T switch. Because there are so many ways to do this, I'll leave the final "imagineering" to you.

For the PRO-2006 the pins are 7-10. Other scanners will require you to look in your service manual's schematic for the combination.

ANTENNA ATTENUATOR

If you live in an area where there are lots of too strong signals, this gadget might be the solution to limiting overmodulation in your scanner. The nice feature of this attenuator is its flexibility. As each "bank" is switched in, the attenuation for that bank is added in so the combinations allow you to fine tune the signal to fit your needs. You'll have to do some hunting for the resistors but other than that, everything else is easy. Ideally, all the resistors should be as close to value as possible. The trick is to buy a few of each and use your meter to check. Start with 5% tolerance. Use 1/4 watt resistors, and do not try to transmit through this attenuator, it isn't strong enough to take that kind of load. Remember, this will attenuate *all* signals.

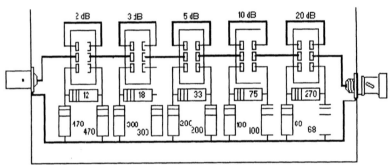

Building it is a piece of cake: each segment is the same as the others, except for the resistor values. The easiest way is to locate the smallest sized <u>metal</u> box you can work with, mount the BNC connectors and the DPDT switches, and then you can work directly on the switches to add the other components.

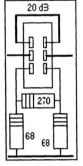

To preserve as much signal as possible, keep the leads as short as possible. A variation you can try is at the output end. Rather than a BNC connector, use the coax that will go to your scanner. This will eliminate one connector, and save a couple of bucks as well! In the insert are the specs for adding a 20 dB segment. Actually, you can add as many segments as you want, and switch things in and out as needed.

ANTENNA TUNER

This is a "luxury item" for scanner users and may not help your antenna system at all. On the other hand, after you've done all the other stuff, you might build this just to see what it does! In order to properly evaluate whatever results you might get, a sensitive meter is a necessity, so you can compare before and after results. It has many variables, and takes a long, fussy time to adjust. If I haven't talked you out of it yet, let's get started!

The variable capacitors can have either single- or double-spaced plates, and for scanners the 365 pF is probably the best choice. Prowl around your electronic surplus store, and you should find a couple for about a buck each. The type is not critical as long as they are alike and can be easily mounted. Use a plastic pill bottle for the coil form, and it is OK to use the bottle in place for mounting purposes. Start winding the coil by making the 2 ½" by ½" turns, and then wind 25 turns on the pill bottle adding the taps as shown. In a coil of this type, the spacing between turns should be the same as the diameter of the wire used. #14 gauge wire can be used for all the wiring, but keep the leads as short as possible.

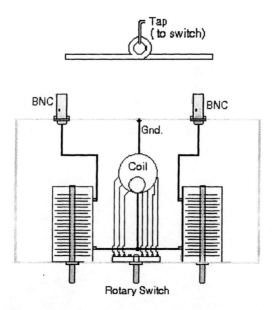

Rotary Switch

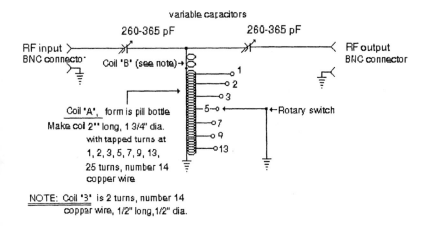

NOTE: Coil 'B' is 2 turns, number 14 copper wire, 1/2" long,1/2" dia.

Tuning up: With the tuner out of the circuit, choose a known weak station, and check your meter. Now connect the tuner with the variable caps set at midrange, and the rotary switch at the low (tap 13) position. Check the meter reading again. Watch the meter as you adjust one of the variable caps and see if the signal rises. (At this point, use the same capacitor for all settings, while leaving the other one alone.) Repeat this as you rotate to each switch position. Hopefully, on one of the taps you'll see a marked rise in signal. When (and if) this occurs, use this tap for the rest of the adjusting. Peak the variable cap, and then adjust the other variable cap for any further peaking of signal. If you are a short wave listener, you can use this tuner to good advantage with a "long wire" antenna, by connecting the wire to the primary output lead.

MULTI-PURPOSE ANTENNA TUNER

This tuner is actually for transmitting, but thanks to the "law of reciprocity," it can be used as well for scanners. Unless you have a good, sensitive meter on your scanner, you will not be able to use it, since the only way you can tell if you are improving your receiving antenna's match is by watching the meter while you "diddle" knobs with a station of known value, preferably a weak one...

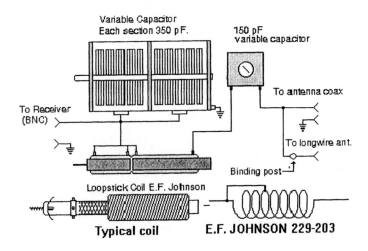

This circuit is best built in a metal box, to shield against stray RF signals, and any long runs should be of ground shielded coax. I keep a couple of yards of RG 174 coax as hookup wire for use with any antenna related project. The diameter is small and it's easy to use. The frequencies at which scanners excel are delicate things, and are best kept "inside the wire!"

Note: While any 350-400 tunable coil can be used, The Johnson rotary inductor (229-203) will add more fine "tunability" and looks really dramatic! Prowl the electronic surplus houses for one. Once you've peaked this tuner to your antenna, it will do the best job right at your antenna. If you go this far, make sure the box is separated from the antenna's metal parts, by the diameter of the coil at a minimum. Remember, this tuner is a "luxury item," and just for the absolute purist!

IMPEDANCE MATCHER

Almost never does an antenna's impedance match the impedance of the cable! Here's an easy to make device that will do the job well and simply. It's actually a matching transformer called a BALUN. This is an abbreviation taken from BALanced/UNbalanced, and that's what it does. It has to be installed right at the feed point of your antenna, and depending on your antenna-coax combination, it can make a big difference in the quality and strength of the signal you receive. (It may make no difference at all that you can tell, but that only means that you were near right in your selection.)

Calculate the length of the balun loop as 1/4 wavelength of the low end of your primary frequencies. Take care to seal the ends of the coax with a silicone sealer to keep Ma Nature at bay, and support the loop.

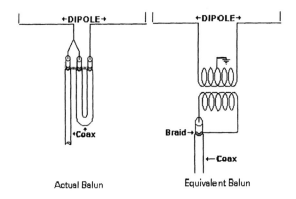

Actual Balun   Equivalent Balun

EASY SCANNER AMPLIFIER

The Archer (# 15-1607) portable TV amplifier could have been made for scanner users! It's very low-noise, and the range is roughly 50 to 900 MHz, suiting a scanner's needs nicely. The power is provided by 2 built-in AA cells, and included in the price is a separate AC power supply.

The antenna is a "rabbit-ears" dipole, with some directivity if that's your choice. The size is small enough to Velcro it right to your handheld, and it comes with a nice built-in multi-position stand, so you can set it along side your base for an antenna which, because of the contained amp, will be far superior to the little whip that comes with your scanner. There are a couple of ways to adapt it to scanner use, the

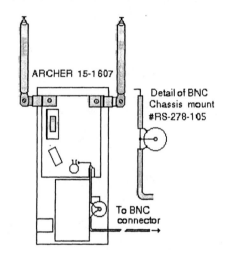

simplest being cutting off the original plug and replacing it with a male BNC or whatever matches your scanner. A little more elaborate work will give you an in-line amplifier that can be used to "pre-amp" your base antenna.

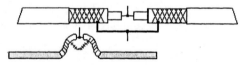

As shown in the drawings, there is just enough room to fit a Radio Shack #278-105 chassis-mount jack on the right side (as you look at the open back).

To open the case, you'll need to remove the plastic data plate in the center of the back (there goes the warranty!) and you'll find the Phillips-screw that holds things together. Remove that first and then, from the bottom up, pry open the case (it comes hard!). Once you have the case open, you will see the antenna lead and be able to choose where you want the chassis plug to be located.

The easiest way to make the connection is to strip off a little of the outside covering of the "cable" to expose the inner braid, which will be connected to the ground side of the chassis connector. Work carefully to avoid nicking the braid. "Kink" the braid double to expose the inner conductor, and without cutting the braid, carefully strip the insulation from the inner conductor. Solder the inner conductor to the center terminal of the chassis connector, and the braid to the outer. Note: the wire used in the amp is very delicate, so take your time and be gentle!

Now, button things up, from the antenna end down. You may find the bottom-most battery spring is in your way for closing the case. Just flip it inside the case with a thin-bladed knife, or similar tool. Replace the center screw and enjoy! Extend both elements fully and orient them vertically for normal use. You will notice a slight directional effect with both elements up as you rotate the antenna. Also, try using the antenna with one element measured to half the length. At that point the amp works like an end-fed J-pole antenna. If you are a heavy 800 MHz user, the best antenna length is 14".

SCANNER AMPLIFIER

Try this simple little amplifier on your handheld scanner for a bigger sound. It isn't sophisticated by any means, but it will drive a 4 ohm speaker (or the same value as the speaker in your radio) really well. The thing is very easy to build, in fact you don't even need a perfboard to work on. All the components can be wired point to point on the back of the variable pot if you choose. Use a mini-box just big enough to hold the battery, and because it is self powered, you can use it for a variety of purposes. Further, if you wish, you can build a speaker right into the case, using the headphone jack leads, or leave it as is and plug the speaker into the jack.

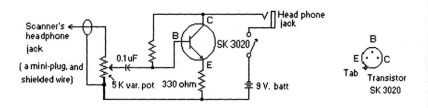

With a little "imagineering" and the basic circuit, you can come up with lots of variations. I'm not certain, but it's likely you can take the voltage from your scanner and build this circuit in the case, although the drain will take away a lot of your "scanning time." Use a variable potentiometer with built-in switch to save space.

NOISE LIMITER

If your area is cursed with noises other than the frequencies you want to hear, you'll find this a handy gadget. All this does is reduce excessive signal levels caused by a variety of sources. When it's working, you won't know it, because it take out sudden, high level, bursts of noise in microseconds and leaves you with a nice, clear signal. The easiest way to build it is to start with a 2-cell battery holder (AAs will last for nearly their whole shelf life). Pay lots of attention to the polarity of the diodes, or the thing won't work at all, and mark the position of the switch so you know if it's on or off.

Tip: Whenever you are working with "store bought" diodes, it's a good idea to check 'em out first. Most of them are OK, but too often, a bad one sneaks in. They're easy to check, just put them across the leads of an ohmmeter, and read them both ways. The diode, if it's good, will conduct in one direction, and not in the other. The potentiometer will allow you a lot of range to play with, and you'll easily be able to adjust it to a level that takes out the noise (mostly) and leaves the signal.

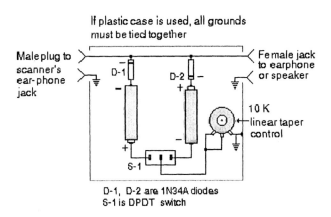

D-1, D-2 are 1N34A diodes
S-1 is DPDT switch

NICAD "FIXER"

Don't toss that nicad cell in the trash until you try this circuit. A nickel-cadmium battery can often appear dead, when in fact it's internally shorted by web-like tendrils that grow inside cells you haven't used for a while. These shorts gradually discharge the cell and prevent it from being recharged. These "whiskers" can be blown away with a high current applied for a few seconds, allowing the cell to be recharged normally. This circuit can be built to accommodate any size cells as long as all the cells are the same size. You need two fully charged nicads to give sufficiently high current to get the effect. Size AA cells can deliver a continuous current of 0.5 amp, plenty to do the job. When you close S-1, a shot of current goes through the "bewhiskered" cell and causes the tiny shorts to open, and as a fringe benefit, if the cell has become reverse polarized, the polarity is corrected. You can build this circuit for just a couple of bucks (you knew I was going to say that!), especially if you have an "S" meter lying around. Use the appropriate sized battery holders for your needs, and if you set up in a configuration with the good cells in one holder it's easy to keep track of what's good and what isn't.

S-2 and the light bulb aren't really required, except to let you see the state of things, and the meter, being part of the circuit, eliminates the need for an "external" meter. To determine the value of R-1, place a 50K pot in the circuit and adjust it until the meter gives full deflection, and then measure the value of the pot, and use a fixed resistor of that value in the circuit. (Another old guy's trick!)

How to use it: With two fully charged nicads in the main cell holder the meter will read "0" until S-1 is closed. (When you close it you should see 2.8 v) Put the "dead" cell in the fixer holder. There should be no reading. If there is reverse polarity, the meter will show negative. If the cell shows a positive reading you *should* be able to charge it normally. If it shows completely dead, press S-1 for a few seconds (count three) and watch the meter rise. If there's no reading the battery is too far gone to save. If you don't get 1.4 v or close, try S-1 again to see if it brings up the voltage. If you are successful, a couple of regular, full discharge/recharge cycles should restore the battery to as good as new.

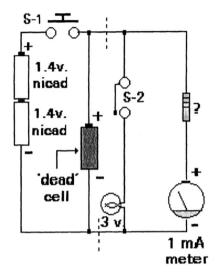

Note: S-1 is momentary contact, the flashlight bulb is 3 v and is a load for the meter, S-2 is to monitor the operation and should be a toggle or slide switch. The bulb may or may not glow. Some will, some won't. The circuit right of the dotted lines is just to monitor the action and is not necessary except for that purpose.

NICAD CHARGER FOR AUTOMOTIVE USE

With a handful of diodes and very little else you can, with this gadget, charge your scanner's battery pack on the run! If you're on a camping trip, for example, this is a way to keep your batt-pac hot when you're out in the "boonies." You'll need to do a little figuring to find the correct number of diodes to make the "voltage regulator" part of the circuit, but even for me this is easy... Check the total output of your battery pack, and figure each diode to drop 0.7 volt. For example, if your pack were rated at 10.9 volts, it would require 10.9 / (divided by) 0.7 = 15.57, which is close enough to 16 diodes.

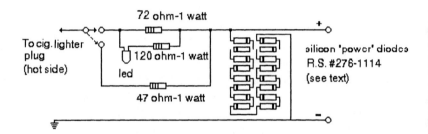

Look at the schematic for the series wiring hookup and while you're there, note that the part of the circuit that goes through the 47 ohm resistor is for fast (5 hours) charging. I suggest you only use this for emergency needs, as the rate is high, and could eventually cause harm to those expensive nicads. The normal charge takes about 10 hours for a full charge. The LED can be nearly any one out of your junk box. If it looks too dim, or too bright, the 120 ohm resistor can be changed to suit, using a resistor value from 47 ohms on up.

17

Antennas

ANTENNAS ARE MODS TOO!

A scanner can only do so much. Modern scanners are efficient receivers as they come from the box, and internal "diddling" requires skill and equipment that are not in the realm of ordinary modifying. (This means me!) However, there are many possibilities for improvement using antennas and good cable. This chapter presents some homebrew antennas, typically low cost with simple tools. The results of a good antenna system can be astonishing.

<u>Line of Sight:</u> No matter how efficient your antenna, the scannist's enemy is always line of sight, which can vary enormously with locale. Line of sight, of course, has to do with how far away your antenna can "see" a transmitter. If you are in the flatlands, LOS might be limited only by the power output of the transmitter you are trying to hear. The same thing applies if you are on a mountain peak. Most of us, however, encounter limits. Hills, rivers, streams, RR tracks, buildings, and power lines, even street orientation, can and do limit a scanner's range. Even the operating frequencies you are trying to receive offer obstacles. (Lower frequencies travel farther.)

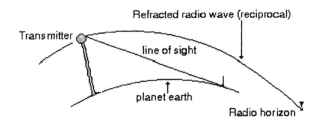

How to cope: Raise the antenna. With a given antenna, the higher it is, the farther it can "see." Even 10 feet more can yield a dramatic increase in range, and up to a point, higher is better. However, the higher it is, the longer the coax, and long runs of coax are limiting. Wind load is a serious factor, too, so 40 feet is about the maximum practical height. If you opt for a mast longer than 10 feet (and even then) mind the danger of power lines; use too much guying versus too little. When you install an antenna, seal all the connections, with boots if available and silicon sealer, always. A tiny bit of moisture in a connector, and/or in the coax can ruin all your work. Even in ideal locations, at least once a year check your connections and "de-corrode" your antennas with steel wool. It can make a profound difference in getting that delicate signal to the scanner.

Too often, the scannist can't use a tower or mast. Neighborhoods have antenna restrictions as do apartments and condos. If you  are in that category, there are still ways to "beat the system," at least to a point. In this section, you'll discover ways to do that.

There is a manufacturer who makes an antenna disguised as a roof vent, and I heard it does a good job. Be a patriot! (and hide your antenna in or as a flagpole.) Put the antenna out on the balcony or patio, especially if you live in a metal framed home. Many homes in California, for example are stucco, and the stucco is hung on "chicken wire" which makes a nearly perfect Faraday shield, usually with an excellent ground. (Bad news for a scanner!)

Use an outdoor antenna *indoors*. I have a discone in one corner of my second floor workshop, and with it I've received a powerful repeater from a little over a hundred miles away.

Life after Cellular Mods: The letters I see cause me to think that many readers believe the only thing that can be done to modern scanners is to open the illegal cellular phone frequencies, and that's the end! Not so, my friends, not so. For the most part cellular gets boring pretty quick. (There are exceptions!)

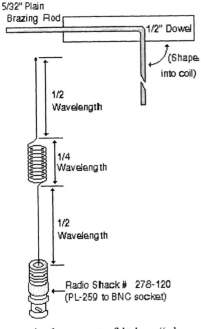

5/32" Plain Brazing Rod

1/2" Dowel

(Shape into coil)

1/2 Wavelength

1/4 Wavelength

1/2 Wavelength

Radio Shack # 278-120 (PL-259 to BNC socket)

For me, the most fun comes from enhancing a scanner beyond what comes in the box, and happily, there are ways to do that, often without removing a screw (and the warranty).

Because I hate math, and love antennas, I've spent years finding simple, easy ways to do what seems like endless calculations for antenna design. What I found out, and want to share with you, is that most of it is a "piece o' cake!" (as long as the batteries work in my calculator). These tips are all you need to design ordinary antennas!

Definitions:

L = length   WL = Wavelength
V = velocity factor   F = frequency MHz

F (frequency): You need to know the band or sub-band you want to receive. This might be UHF, VHF or other, or a combination of bands. Generally, your figures should reflect the lowest frequency you're interested in, since the lower the frequency, the longer the antenna. All scanner frequencies are normally expressed in **MHz.**

WL (wavelength): For our purposes wavelength and F MHz are the same thing; knowing either one is the same as knowing the other.

V (velocity factor):   All coaxial cable shows resistance and capacitance (resonance) to an antenna circuit. Usually, the V is expressed as a percentage, so if you know the V factor (from a chart) it's a simple thing to work on the calculator. The velocity factor affects the length of the antenna and the cable, and will make both shorter. If you lack the V factor, an easy ball park figure to work with is: bare wire = 95%... insulated wire = 85%. This won't be right on, but scanner antennas are very forgiving of errors. (Thank heavens!) Believe it or not, you now know everything you need to design an antenna, except a few divisors.

My favorite shortcut divisor is 2834/F MHz = 1/4 WL in inches... For example, 2834 divided by 150.000 MHz gives 18.92 inches (a quarter wave). Multiply by 2 for a half wave (37.84) or by 4 for a full wave of 75.68 inches. This stuff is easy! The shortcut divisor has a bare wire V-factor built in to make things really simple.

Here are a few divisors for fussier work. Let's assume you're building a yagi (beam) antenna with one reflector, one director, and one driven element. Here are the divisors:

Driven element: L= 11826 / F   V = full wavelength with velocity factor.
Reflector: L= 12562 / F MHz V = full length with velocity factor.
Director: L= 11248 / F MHz V = full length with velocity factor.

Note: The easy way to "ball park" the director and reflector is to add or subtract 5% respectively.

Now, we can build one (at least in theory). A few seconds' study shows that everything you need (almost) to build any antenna is here. All we have left to "figger" is the spacing between the elements. It's called **K** and is supposed to be a constant, but it turns out to be a constant that is affected by climate, temperature, and geography. Some constant, huh? Actually, in the great USA, about 20% WL works in almost all cases for a receiving antenna.

Hams note: The antennas I write about are not intended for transmitting, although in many cases they would work. These are scanner antennas, and this is a scanner book!

The law of reciprocity: Antennas transmit and receive equally well on given frequencies, but if you are a ham or CB'er with a zillion watt power supply, some of these antennas I write about will melt! They are primarily built with receiving in mind.

COAXIAL CABLE

In the simplest form there are three basic elements in a scanner system: the scanner, the antenna system, and the connecting cable. Given any specific scanner, the other two factors can provide nearly endless combinations, some of which can cause vast improvements in reception. Each scanner system is different, primarily affected by location and antenna height as well as the link between the two, the cable. Every coaxial cable is made up of at least four parts:

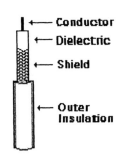

1. The center (primary) conductor: May be solid or stranded wire, normally 18 or 20 gauge, but can be from 13 to 26 gauge.
2. Inner insulation. May be foam, nylon, Teflon, or other.
3. Shield: May be copper braid, copper spiral, aluminum braid, tinned copper braid, spiral foil, or bonded foil.
4. Outer insulation. May be rubber, nylon, polypropylene, or other material.

As you can see, we've opened a can of worms. Now to narrow the limits a little. I can only define some of these things in a general way, based mostly on experience, mine and that of others.

1. Stranded wire is for the most part more flexible, easier to use, and less vulnerable to kinking and nicking. Nicking is caused by careless stripping, and will always end with a broken wire in your future. It can also result in losing some of that precious signal you're trying to get into your scanner. If you nick a wire, always cut the end back and start over! Stranded wire is slightly more difficult to use with some of the connectors, but since it conducts better, is worth the effort. Aluminum wire tends to become brittle with age, and is way down on my list. Wire diameter needs to be factored in mating connectors to the coax, but in normal wire runs of 50 feet or so the difference in conductivity between 18 and 20 gauge is not great. (Lower number equals larger size, and bigger IS better!)
2. The most common among these is nylon, but foam core insulation has the edge in reducing line loss.

3. On the very bottom of the list for shielding is spiral wrapped copper, and only slightly better is any class of braided wire. Spiral foil is a somewhat better choice. Bonded foil gets my top rating.

4. The outer insulation is all more or less the same regarding a normal coax installation. Some outer coverings can be far more durable than others, and more or less vulnerable to weather and ultra-violet light, but other than the real "brand x's" should give a two year life, unless you operate under extreme conditions.

Here are some standard types and sizes of coaxial cable in the 50 ohm range to show the varieties:

Mouser Stock no.	type	Conductor AWG	strand	%	of shield type	Impd	od(in)
RG8 Cable							
1109-38	bcu	13	7x21	70%	copper braid	50Ω	.405
1180-38	bcu	9.5	solid	100%	bonded foil	50Ω	.405
1108-18	bcu	6	19x28	95%	copper braid	50Ω	.242
1154-12	bcu	13	7x21	95%	copper braid	50Ω	.405
1198-12	bcu	11	19x24	95%	copper braid	50Ω	.405
1107-12	bcu	16	19x28	70%	copper braid	50Ω	.242
RG-58 Cable							
1116-18	bcu	20	solid	70%	tinned copper braid	55Ω	.195
5045-18	bcu	20	solid	95%	tinned copper braid	55Ω	.195
1188-18	tcu	20	19x.007	95%	tinned copper braid	50Ω	.195
1140-18	bcu	20	19x32	95%	tinned copper braid	50Ω	.178
1166-12	bcu	20	solid	95%	tinned copper braid	53Ω	.195
1166-18	bcu	20	solid	95%	tinned copper braid	53Ω	.195
1178-12	bcu	20	19x007	95%	tinned copper braid	50Ω	.195
RG-174 Cable							
1156-12	bcw	26	7x3	95%	tinned copper braid	50	.100
1156-18	bcw	26	7x34	95%	tinned copper braid		.100
LAN Cable							
5779-18	bcu	20	19x32	77%	tinned copper braid	50Ω	.182
				100%	bonded foil		

BCU Bare copper
BCW Bare copper weld

Impedance: The ideal impedance of the cable chosen should be, nominally 50 ohms to fit the design parameters set for scanner use. Impedance can be defined as a combination of reactance and resistance, but to simplify things to the lowest common denominator, (so I can understand it!) it can simply be thought of as resistance in the coaxial line. (Electronics engineers are cringing!) For our purposes here it is sufficient to say that the coax you buy should be as close to 50 ohms as you can get, rather than 75 or 300 ohm coax.

Note: The above material was taken from MOUSER Electronics' Catalog and is CAROL cable, 1-(800)- 34 MOUSER.

Velocity factor: The next step to a "perfect" system is the consideration of velocity factor, the effect of the insulating material itself. Each type of line has a specific velocity factor and can make a difference to your scanner's function. The velocity factor can, depending on the formula you use, also affect the element length of home built antenna systems. The math is simple, if you have the velocity factor of a given cable: For antenna elements the "rule of thumb" is bare wire = 95%, insulated wire 85%. For receivers, this will be close enough for any practical purpose. 246/F MHz = full wave length in feet × velocity factor (%) = actual length. Following are some typical coaxial factors for your use. (The coax chosen are in the 50 ohm range.)

TYPE	IMPD	VEL%	DIELECTRIC	TYPE	IMPD	VEL%	DIELECTRIC
RG-8X	52Ω	75	Foam PE	RG-58C	50Ω	66	PE
RG-8	52Ω	66	PE	RG-141	50Ω	70	PTFE
RG-8 Foam	50Ω	80	Foam PE	RG-141A	50Ω	70	PTFE
RG-8A	52Ω	66	PE	RG-142	50Ω	70	PTFE
RG-9	51Ω	66	PE	RG-142A	50Ω	70	PTFE
RG-9A	51Ω	66	PE	RG-142B	50Ω	70	PTFE
RG-9B	50Ω	66	PE	RG-174	50Ω	66	PE
RG-17	52Ω	66	PE	RG-213	50Ω	66	PE
RG-17A	52Ω	66	PE	RG-214	50Ω	66	PE(2 Shields)
RG-55	53Ω	66	PE	RG-215	50Ω	66	PE
RG-55A	50Ω	66	PE	RG-223	50Ω	66	PE(2 Shields)
RG-55B	53Ω	66	PE	9913	50Ω	89	PE(AIR SPACE 2 Shields)
RG-58 Foam	53Ω	79	PE				
RG-58A	53Ω	66	PE	9914	50Ω	78	PE (Foam 2 Shields)
RG-58B	53Ω	66	PE				

Dielectrics:

PE	*= Polyethylene*
PE Foam	*= Polyethylene Foam*
PTFE	*= Polytetraflouroethylene*
	(Teflon)

ANTENNA TYPES

For the purpose of scanning, antennas are intended to receive signals. Any given antenna is reciprocal, which simply means that it will transmit as well as receive, but as a receiver it does not need to be built "heavy" to handle the types of power requirements a transmitting antenna might. In other places in this book you'll find some of these ideas repeated, but here's a general overview to clarify some fundamentals.

When building an antenna to receive a group of frequencies, choose a frequency at or close to the bottom of your selected band. Since antenna length increases as frequency decreases, doing the math from the low end will usually result in an antenna that will best suit your needs. Within reason, choose a larger diameter for a radiator (element) over a smaller one when building an antenna for strength, but also because a larger diameter radiator will increase the bandwidth at the frequencies you're trying to capture.

There is no way at normal scanner frequencies to defeat the line-of-sight (LOS) rule. Actually, given optimum conditions you might "see" LOS plus 30% at an absolute maximum. Within your pattern, another factor is the power output of the transmitter you're receiving. If that station is weak, the best results are obtained by using an antenna with directional capability, so you can point your antenna directly at that station. Alas, if you are pointed at a given station, and your antenna is well made, other signals at the sides and back of the antenna will be reduced. Matching coaxial line to your system can make some difference, but if your cable run is short (25 feet or so), the more exotic cables probably won't make much difference, except to your wallet (or purse!).

Same for amplifiers. I get letters from new scanner owners asking what kind of amplifier they need to get signals from two states away. The answer goes back to LOS. Often an amplifier will do more harm than good by increasing the noise level, or by bringing you strong stations that you already receive, while overloading your scanner to the point of uselessness. Sadly, the only way to find out if an amplifier will be effective is for you is to buy (or build) one, and try it.

Selectivity and sensitivity are not magic words! If your scanner is well made, and has proper shielding and filtering, the numbers make very little difference. I'm a strong believer in the empiric method: If it works well the sensitivity and selectivity are right!

Sensitivity: Simply defines a receiver's ability to hear the weaker signals. The confusion stems from the uses of the definition. One view relates to the ability of an antenna to "see" a specific output level, and the other is the ability of the receiver to separate signal from noise. Extraordinary claims for high sensitivity should cause you to be suspicious.

Selectivity: A measure of a receiver's ability to reject signals other than the signals the receiver is intended to hear. Selectivity is related to bandwidth and a radio with low selectivity is likely to receive images in place of the actual signals desired.

Dials and meters: I love 'em! However, the usefulness is limited to comparing one antenna to another, or on the rare occasion when you decide to re-align your scanner, and if you are qualified to do that you already own a meter and an oscilloscope that will do a better job. (Letters again: I get comments telling me that with meters you can tell if a station is near or far away, or if a mobile is coming closer. Maybe, but what purpose is served? (Just for the record, there are three meters connected to my venerable PRO- 2006!)

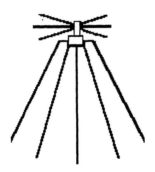

DISCONE ANTENNA
omnidirectional
[receives equally well in all directions,
usually with a very wide band coverage]

Antenna types: The standard for scanners is probably the Discone. Essentially a dipole with a top hat capacitor, the frequency range is 25 MHz to 1300 MHz (1.3 GHz.) There is a tendency for weak signals at the lowest end, but there's a simple mod to fix it elsewhere in this book. The radiation (reception for scanners) pattern is the same "doughnut" as any regular dipole.

Vertical dipole: This is the fundamental antenna from which all the other scanner antennas are derived. By the way, often antenna manufacturers use an isotopic antenna to evaluate the dB figures for an antenna. Since an isotopic antenna is just an ideal point in space, this figure is misleading in the real world...

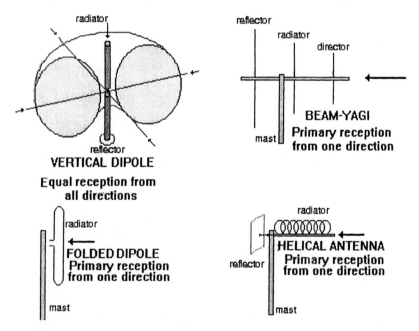

Folded dipole: A simple way to introduce gain into an antenna system. (More about gain elsewhere.) Some directivity.

Beam-Yagi: A simple way to make an antenna directional. It's nice to use if you can either point it with a rotor, or if all your signals come from one direction. Signals at the back and sides of any directional antenna are reduced or lost. Considerable gain.

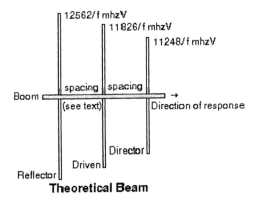

Theoretical Beam

Helical antenna: This is the "shotgun" of common antenna systems. Very directional and a gain figure of 14 dB is common.

Log-periodic: (Not shown here) Slightly more difficult to make, the directivity and gain can be very high.

Decibel-Gain: As it relates to antennas, Decibel and Gain figures can be thought of as the same thing. A Decibel (named after Alexander Graham Bell) is simply a way to define loudness. The human ear hears louder sounds in a logarithmic way, so units of dB are measured the same way. For scanner use the math really isn't required, so it is not included here. A change of one dB is defined as the smallest detectable increase in loudness under ideal conditions. Common logarithms (base 10) are used.

Balun: The term Balun is a contraction of *balance to unbalance* and that's what it does. Even though it doesn't look like it, a balun is simply a transformer with the ability to equalize the differences between an unbalanced coaxial cable and a balanced dipole or, in some cases, to equalize both. Once you see the schematic view it's simple to see how it works. A balun is cut to a division of a wavelength to cause a match between the coax and the antenna when the impedances are unequal. A 1/4 wave length equals a 4:1 ratio, the figure most used in scanner antennas. Most center fed antennas are not improved with a balun and the equalization is only theoretical, but if an antenna is end fed a balun can make a big difference. There are other types of baluns and matching transformers but the one shown is the most simple and common.

Grounding: Except for a minor safety margin and a little possible noise reduction, grounding a modern scanner system is unnecessary, Unlike transmitting systems where grounding contributes greatly to the efficiency, very little improvement is shown at scanner-received frequencies. The benefits that can be derived result from reduced static build-up, which might reduce the chance of being visited by Mr. Lightning! The bottom line (for me) is, if it is easy to ground your system do it, but if it isn't, don't worry. If you do choose to ground, a three foot ground rod driven almost to earth level with an aluminum cable to your mast is all you'll need. The best lightning protection is to disconnect your antenna and unplug your scanner when a lightning storm is imminent, but that's when the scanner is likely to be working on exciting stuff! (Oh, well.)

SAFETY FOR YOUR ANTENNA MAST

There you are with your scanner, coax, and a shiny new super-duper DC to LIGHT antenna and you can't wait to use it to scratch the sky! Please spend a couple of minutes to read this first. Modern hardware makes it easy to install a mast. You can buy "push-ups" to 36 feet, and other mast combinations will allow you to practically poke the clouds. The higher you can raise your antenna the better (until the evil Mr. Cable Loss begins to take his toll) but physics and Murphy's law dictate that unless you put that mast up right, it WILL find the best route to your neighbor's greenhouse. Or worse.

Planning: Take a long look around at your options, and then look again. See where the obstructions are, the power cables, and the location of your feed-in for the scanner. You might be able to "piggy back" on your TV antenna, and even use the rotor to direct a beam. Nice! If you need to bring that mast down, make sure when it goes back up that it's properly guyed. The usual antenna used for scanning is light weight, and not very big, but at the top of a mast it becomes the heavy end of a lever, and just a little wind can exert incredible forces. Ice can add a lot of weight, even to a small antenna. Always over-design your antenna mounts.

Mounting: There are several mounts available to choose from and a catalog will help you make a selection. Wall mounts of several types are a choice and roof top mounts, including chimney mounts, offer even more options. (However you mount an antenna, make sure the anchor point is solid; an antenna can pull out lag screws in minutes, or crumble an old badly mortared chimney with ease.)

Guying is the most important factor. Whatever your choice of guy anchors, they must be secure. In a roof, for example, use the roof joists (beams) rather than just the roof surface. By the way, coat with roofing tar anything driven into the roof, not only to prevent leaks, but also to act as an anti-corrosive. Never guy out an antenna at less than 1/3 the height. Even this is cutting it fine, but it is adequate.

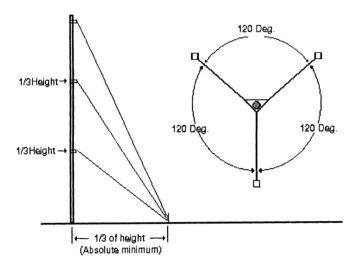

Use turnbuckles to adjust your guy wires; every guy wire stretches and turnbuckles will allow you to make adjustments. Use the proper size wire clamps, at least two to a guy wire. No matter how you twist and knot a guy wire, it will work loose at the least opportunity, and knotting weakens the wire.

Choice of guy wire: This is a never-ending argument, as far as I've seen, with all the "arguers" extolling the virtues of their choice. My overall favorite is stranded aluminum.

Expect your guy wires to stretch. Once you have the antenna up, the stresses that will occur from wind, and the tensions put on the guys, will quickly introduce a surprising amount of slack, so after you've had it up for a day or so, go back and re-tension the turnbuckles.

Use stand-off insulators to keep your coax from rubbing against the mast or the structure on which the antenna is mounted. It only requires a little friction to rub through and ruin that expensive cable. Avoid any kinks in the coax. Losing the integrity of the "braid" or the dielectric will cost you signal.

Don't use splitters or adapters. It is cheaper and better to install the proper connector on the cable before you "hook-up." Any compromise you make on the straight cable will eat your signal. The whole point of the antenna system is to bring the maximum amount of signal to your scanner, and although you may never know you are losing part of that signal, it's to your advantage to do everything you can to install a perfect system. Get the best coax you can; 50 ohm cable is the ideal to fit the design of your scanner, and an impedance higher than 53 ohms will cause significant loss. Trying to use other lead-ins with transformers adds cost and loses signal!

Minimize length of coax cable: Where you can, keep the coax run to a minimum. Amplifiers used for long runs often do more harm than good, by amplifying unwanted noise and static.

Develop an effective ground. Along with providing some lightning protection, a good ground will help reduce static buildup in your antenna, and in most cases improves your signal's clarity.

Be safe. Get the help you need to install the antenna, and use good ladder practices. "If you're going to fall off the ladder, make sure you're on the bottom rung!"

DIPOLE YAGI ANTENNA

The possibilities are great with this easy to build directional antenna. If you have a rotor system, it's easy to make in either a specific frequency grabbing form, or as an all purpose antenna that can be readily pointed at specific targets. If you choose to mount it on a metal mast, stand it off by at least 1/4 wavelength to reduce any effect the mast might have. There's more, though! This antenna can be built light enough to mount right on your handheld scanner, and you can take it with you to an air show, or on a camping trip, or wherever you choose to carry your "ears." The boom can be PVC tubing or any other light-weight material. The folded dipole and reflector can be # 10 or 12 wire, or for more durability, aluminum or copper tubing. This antenna can be fed directly from coax, but a balun or other matching transformer might cause some improvement. This antenna should give a minimum of 6 dB gain.

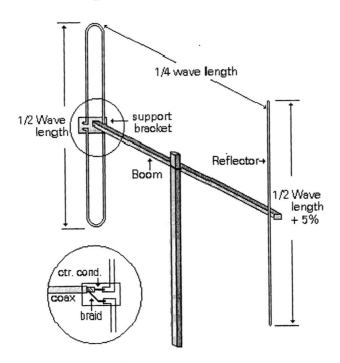

"BAZOOKA" ANTENNA

As far as size is concerned, this antenna is ambitious to say the least! It's also one of the most effective dipoles I've run across. Some of the parts for this one you'll have to "imagineer" to fit. If you have a lathe, making the plugs is easy; if not, poke around your junk box or the hardware store. The outermost tube is cut from 2" light-weight electrical conduit, and the mast is 1" pipe, cut to a length to suit your own purpose for mounting. The coaxial cable is fed through the mast (and the plugs) and the center conductor is soldered to the SO-239's center contact. At the same time, the coax braid is shorted to the mast and the outer conduit.

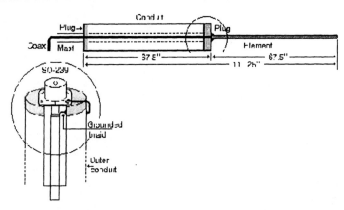

The element is based on a PL-239, and because of the length, has to have a strong physical support. I inserted a piece of brass rod into the PL-239, and then bent it for a tight fit inside a piece of 7/8" copper tubing, split so the tubing would fit over the *insulated outer shell* of the connector. A small automotive hose clamp was used to hold the thing together. (This is not as strong as I'd prefer, but it seems to do the job!) Finish things up by soldering a cap on the top end of the element. The dimensions are for an antenna that works from below 40 MHz well into the 400 MHz band. Using the outer conduit greatly enhances the effectiveness of this antenna. Other antennas can be built for specific bands using the 2834/F method.

DISCONE MODIFICATION

The discone antenna has become the antenna of choice for scanner users. Wide frequency range, omnidirectionality, and ease of installation make this a splendid antenna. Among the choices available, the most popular of the discones is the Radio Shack 20-013. Here are a couple of ways to make it better.

<u>Assembly:</u> The radials have a tendency to loosen from the head casting, possibly due to expansion and contraction. Loose radials can cause serious degrading of efficiency. This is easily corrected by adding proper-sized star washers between the casting and the lock

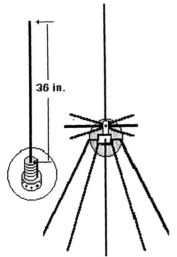

nut on each of the radials, and also the top-hat elements. Don't overtighten the lock nuts, it is possible to pull the radials right out of the casting. The "screw-jack" effect of the nuts can exert enormous pressure.

Although Radio Shack claims a range of 25-1300 MHz, the lower frequency end can be greatly improved by adding a top whip.

On the top of the casting is a black plastic cap. Remove it and expose the threaded stub. Add an automotive replacement antenna to this stub, available at auto parts suppliers for about $5.00. Most are equipped with set screws necessary for good connections.

The antenna is designed to use a PL-259 plug and at the antenna end this is acceptable, but at the scanner end, it's best to use a BNC connector on the cable, rather than an adapter. All adapters cause some signal loss, and the idea is to get as much of that signal into your scanner as possible. Carefully seal your outside connections from the weather, any corrosion or moisture will degrade your antenna's operation. Depending on your mounting, make sure your mast is properly guyed, and be cautious of power lines.

GROUND PLANE ANTENNA

If there's any problem with this antenna, it's more in the mount than in the antenna. Depending on the mast, a bracket is "imagineered" to hang in place as desire. Build the antenna on an SO-239 chassis mount connector. If you have simple shop facilities, the radials can be made with pliers, and held in place with bolts and nuts, or if you are able to do the threading the SO-239 can be tapped and the rod ends can be threaded. I've done it both ways, and other than aesthetics, I saw no difference.

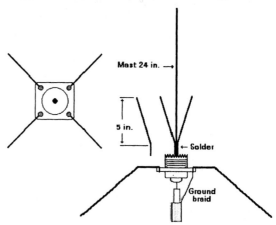

For the element, the formula is 234/F. This formula has the bare wire velocity factor built in. The formula allows you to cut your antenna for specific bands, or for general purpose use.

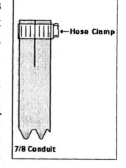

For the radials, the formula is 246/F. Bend the radials down from 30 to 45% for the ground plane effect. The two secondary 5" elements, if used, help a lot with the high (800 MHz) frequencies. The material used for the rods can be any malleable wire rod such as 1/8" brazing or welding rod. Electrical connections need to be heavily soldered or clamped, and well insulated if the antenna is to be outdoors. I have two of these antennas working indoors, one cut for media stations, and one for military aircraft.

See the diagrams for an easy mount. Split a piece of 7/8" conduit, run the coax through it, solder and insulate the connection, drop the connector into the conduit, and tighten the hose clamp.

MILITARY AIR LOG-PERIODIC ANTENNA

This antenna is a little more ambitious than most in this book, but if you are a fan of federal or military air frequencies, and the 220 and 440 MHz ham bands, this is an antenna worth having! Before you build it, though, remember that military air frequencies operate on AM, rather than FM, so you'll need to be able to change modes. Further, if your scanner lacks the 380 to 512 MHz band, this antenna is not for you. (PRO-2006 users, arise!)

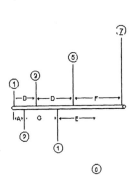

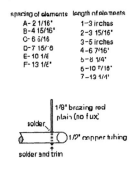

spacing of elements	length of elements
A- 2 1/16"	1-3 irches
B- 4 15/16"	2-3 15/16"
C- 6 5/16	3-5 irches
D- 7 15/16	4-6 7/16"
E- 10 1/8	5-8 1/4"
F- 13 1/8"	6-10 7/16"
	7-13 1/4"

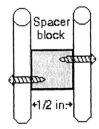

This antenna is really simple to make, although it might not look it at first glance. By making it dual boom, and putting one half out of phase with the other, it becomes very efficient, and by angling the booms as shown, the bandwidth is widened considerably. Use 26 inch long ½" copper tubing for the actual booms, and 1/8" brazing rod for the elements.

Carefully locate the holes, and drill all the way through the copper tube so you can extend the element ends just 1/16" through the copper (see inset). Because of the size of the soldering, a propane torch or heavy-duty soldering gun is needed to solder the joints. Brighten the surfaces to be soldered with a bit of emery cloth, and use rosin flux and rosin core solder. (Acid flux is a no-no!) Make two boom assemblies, and invert one 180 degrees to complete the dipole configuration.

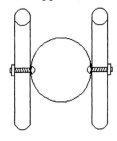

The boom halves must be electrically isolated from each other. At the end with the shortest elements, I used a piece of maple 1/2" square by 3/4" thick, with the screws offset. A block of plastic or any other insulating material can be used as well. Now find the balance point of the antenna to locate the mast. I used a 1 foot long piece of PVC pipe 1¼ " side diameter for a "slip over" mast head so I can use the antenna in other places. Just drill straight across through the boom and mast to locate the holes 1" below the top, and with machine screws, from the inside of the boom, bolt the whole thing together.

This brings us to the back end of the antenna. Locate another 3/4" thick block near the end of the antenna, and pencil in the slight angle and width. Use a wide enough block to make everything solid (3" or so) and to support the balun. The balun is made of a piece of coax 9 1/2" overall with the ends stripped enough to gather together and solder the braid. The wiring is as shown. After everything is wired, insulate and secure the balun, and

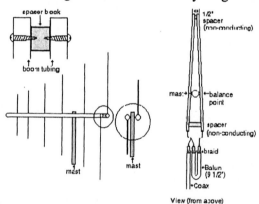

double the coax back to the mast. Because of the way the mast is pivoted, in my case it allows me to give the antenna a slight upward angle that I need to "focus" on the tops of the small mountains surrounding my dwelling, and I use the coax as an azimuth adjustment. For your purposes, you might want to add a small bracket to prevent the pivoting. The antenna gives about 6 dB gain, and covers about 200 - 500 MHz.

STEALTH ANTENNA!

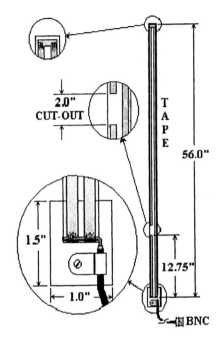

At last there is an antenna you can hide anywhere! Although it uses common enough materials, the application is unique and I think I may have originated the idea. This antenna can be cut to any length to reach into the realm of those stations that you can't quite get, or may be used for general purposes.

First, the math: Here's the part I hate but this time it is so easy even I can handle it! To calculate a quarter wave antenna, just divide 2834 into the frequency in MHz and the dimension comes out the other end in inches. There are more sophisticated formulas, but I find scanners to be very forgiving and this method comes close enough for ordinary purposes.

<u>Note:</u> This antenna is NOT for transmitting. Here's an example for the numbers...

2834/ 150.5000= 16.172". Simple. Multiplying by two gives a half wave antenna of 32.344", close enough to 32 3/8" for our purpose. (The closer you can come to a full wave, the better; however, with the range of most scanners, some of the figures get unwieldy. For example if we pick the center of the CB band, call it 27.5 MHz, a full wave is a little more than 34 feet long!) If you are cutting this antenna to get a particular band, find the "band center" and work with that. For example, if the band you want is from 150 to 180 MHz, split it to 165 MHz and use that figure.

<u>The "Secret" Material:</u> The people who make doll's houses use this stuff every day to wire in the lights and other electrics and any

doll house shop should have it for sale. It's called 2-conductor tape wire and comes in rolls of 5' and 15' length. It is also sold in bulk by the foot, but all dealers may not have it in that form. If they do, of course you can buy it by the foot. It's made by a company called **Cir-Kit Concepts,** and as a last resort you can probably get it from the distributor, 407 14th ST. N.W, Rochester MN 56901. The price over the counter is $7.98 for a 15 foot roll.

At last, here's the good part: Picture a roll of 3/4" cellophane tape with two 1/8" copper strips imbedded in it on the long axis. The tape has a thin, very durable plastic covering and the "back" side is very sticky, covered with a protective strip that is pulled off when you are ready to apply it to the location you choose. For antennas, this means you can hide it behind a picture frame, press it onto a window riser, a door jam, or literally anywhere you can find a vertical surface long enough to accommodate the length of the antenna you make. Among many others I've made are a couple of them on pieces of 1/8" by 1 1/2" trim molding, cut to fit exactly from my floor to ceiling, so I can move them around to where I might be puttering around the house. By the way, this material can be painted to match your decor, which has saved me from comment from an often tolerant wife...

Parts: You will need a roll of this tape, as much coax as is required to run from your scanner to the site of the antenna (possibly 10 feet of RG8), an appropriate connector, usually a male BNC, a small piece of perf board (RS-276-1359A), and a couple of polypropylene cable clamps for strain relief (RS-64-3028).

Cut a piece of the perf board to about 1" × 2", and mount the strain relief clamp (see drawings). For an example antenna, here are the dimensions for a combination 800 MHz (CMT) and general purpose antenna that I have very good results with... Using a 56" length of tape, peel about 3/4" from one end and stick that to one end of the perf board. Measure up from that end 12 3/4", and from ONE side of the foil part of the tape clip out 2" of the foil to make a gap (12 3/4" is real close to a full-wave cellular antenna!). You'll find you can open up the plastic covering of the tape to remove the foil, without altering the strength of the tape. At the top end (farthest from the perf board, peel back just enough of the plastic covering to solder a jumper between the two exposed foil strips. This makes the "long wire" part of the antenna.

Back to the perf board end... You can choose to use a small terminal block if you like for appearance and strength, but it really isn't necessary. The strain relief clamps will be enough to take a lot of abuse. Strip the coax in the ordinary manner, and solder the leads according to your antenna choice. Because scanner receiver antennas are very forgiving, the choice of variables is almost unlimited. In every case, close is good enough. By using the combinations of jumpers and ground you can manipulate the length of the overall antenna to fit almost any need. For example, you might have a window that is too small to take the antenna length you want. Simply add another piece of tape parallel to the first, add another jumper, and effectively "fold" the antenna to reach the length you desire. Remember, normal scanner transmissions are vertically polarized, so you need to find a place to orient your antenna the same way. I have made several of these antennas for a variety of uses, and in many cases, exceeded the capability of my trusty discone, given specific frequency groups. Try this one, It's friendly!

THRU-THE-GLASS ANTENNA

Because of the signal loss, this isn't one of my favorite adaptations, but in an area with strong signals, or when its use is unavoidable, it can be an asset.

Start with a couple of thin 3" x 3" aluminum plates, and cut and bend as shown (a modeler's razor saw does a nice cutting job). Since the ground lead is not used, the connectors you choose must be insulated completely from the plates, and only the center conductor used for connections. Use coax though, the dielectric helps. The version shown is used where you want to keep the connectors intact on your cable, although a better way would be to simply hard-wire the center wire to a solder lug on both the inside and outside plates. The plates can simply be taped to your glass, or for a permanent job, the adhesive used to cement rear view mirrors on automobiles can be used. Double-faced tape leaves too much gap between the plates, and should not be used. The plates make a coupling capacitor, and are not very efficient for transmitting without a trans-match, but for scanner use the limits are more or less acceptable.

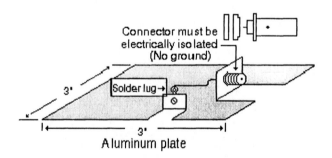

CHEATER ANTENNA FOR HANDHELDS

Make a couple of these to augment your "Rubber Ducky" antenna when it isn't doing the best job. If you are an air show fan, for example, and you can receive the military air band (AM 225 to 400 MHz) this will provide a big boost to your reception. Railroad hobby shops are a good source of brass rods up to 36" long, and piano wire up to 48".

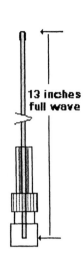

13 inches
full wave

Find the kind of BNC (male) connector that will allow you to feed the antenna rod clear through to the female connection, or pull apart a conventional BNC and modify it to suit your needs. My method runs the antenna element clear through the male into the female side, and then close to the male part of the connection, in order to make a positive connection. Spaghetti tubing is used to insulate the element from the connector, and a larger diameter piece is shrunk around the outside of the connector to make a professional looking antenna.

Math time: Cell phone: 870.000 divided by 2834 = 3.25×4= 13.29" (round to 1/4") for full wave.

Military air: (middle range) 312.50 divided by 2834 = 9.06×2= 18.13" (round to 1/4" for half wave). Other antennas can be cut to suit your needs.

FOLDED DIPOLE ANTENNA

As far as effective antennas go, this one can be made for pennies! Even nicer is that you can roll it up to fit in your pocket and take it with you on your next trip. You need a length of 300Ω twin-lead wire (Radio Shack 15-1158), a couple of washers, the amount of coax you need to reach from your scanner to the antenna, and an appropriate connector, probably a male BNC. Radio Shack sells a prepared BNC cable, with connectors on both ends, 6 foot $5.99, 12 foot $9.99.

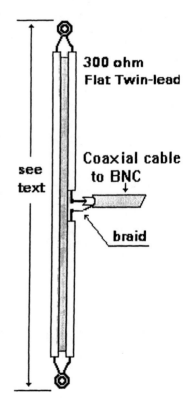

Making the antenna couldn't be easier: Just cut the twin-lead to 48", and at the center point, remove the insulation from one lead for about an inch. Cut this and wire in the coax as shown. Strip 3/4" of the insulation from each end and solder the washers. (The bare wire length, with the washers, makes a hanger and a perfect balun for this antenna.) Surprisingly, you will see about 2 dB gain from this little toy. This antenna can also be made frequency selective by cutting it to the right length using the formula 2834/F MHz = 1/4 wave in inches. If you are selecting a band of frequencies, use the lowest frequency for the math.

"MINIMALIST" ANTENNA

Although this antenna can be made from "almost nuthin," it's an efficient design and will definitely outdo the telescope on the back of your scanner, and probably will do much better than most "rubber ducks"! I used a barrier strip (Radio Shack 274-656) on the test model just to keep the design clean. You can dispense with the barrier strip and use #10 wire for the elements, or to simplify it even further, strip 36½" of outer insulation from a piece of coax (don't nick the braid), and at the 36 inch point, make enough of an opening in the braid to ease the center conductor out through it, along with the dielectric. Then strip the dielectric, again without nicks, to leave 36" of bare wire, and then pull and stretch the braid to its full length, and cut it to 18". As with all scanner antennas, hang it vertically.

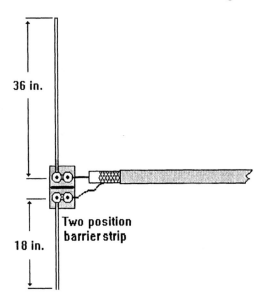

TOP HAT CAPACITORS

An easily made "top-hat" will improve the operation of any base or center loaded antenna. It works two ways; the coil resistance is lowered and the inductance creates a better match to your coax cable. A solid metal disk can be used, but you can also use 10 gauge copper wire, bent around a form to give you about a 6" circle. A small piece of split tubing makes the "hub" and a set-screw or small hose clamp secures the top had in place at the top of your mast. There isn't anything fussy about the construction of this project, and it can be a fun soldering job. Do take the time to make good solder joints. Remember it has to stand a certain amount of wind load if you use it mobile.

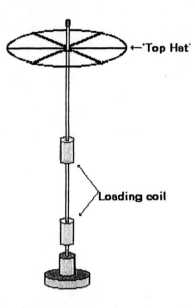

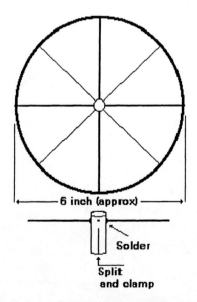

800 MHz HANDHELD ANTENNA

Live in a trunk zone? Enjoy those "middle 800" freqs? Here's an easy to make antenna to go on your handheld scanner that will get you as much signal as there is in your area!

Here comes the math again. (Gasp!) This time though, it is even easier than usual:

2834/ F MHz = 1/4 wave length in inches
2834/820.000 MHz = 3.456'' (1/4 wave length), × 2 = 6.912" (1/2 wave length)

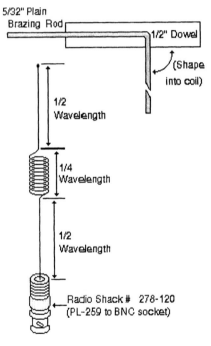

Thanks to the wide spread in scanners, 3.5'' and 7.0'' will work nicely for an antenna that covers close to 800-900 MHz.

The primary element is made from a length of 1/8" unfluxed brazing rod, easy to get from any welding supply store, and from many hardware stores. The easiest way to make the stabilizing coil (it's also a center loading coil) is to make a mark off 8" and 11½" on the rod, and then heat the area of the coil to red with a propane torch and quench it. Brazing rod, when heated close to red color, can be readily wrapped around a piece of 1/2" dowel after making a 90 degree bend (see inset).

From the 8" mark to the 11½" mark, wind a coil with spacing as close to 1/8" as you can get it. At the 11½" mark, bend the rod back into a straight line, adjust the alignment and the coil, and then reheat the coil area and allow it to air cool.

Measure the length of the coil back to the short end (7'') and cut off the inch or so, and then measure the other end (7'') and cut. If everything is right, the overall length should be 14'' plus the length of the coil. With emery cloth, carefully clean an end of the element, insert it into the socket on the connector, and solder it in place.

If you choose, you can dip the whole assembly in liquid rubber (from auto shops, used to protect and insulate tool handles). This same antenna principle can be used with other UHF antennas, although they get long! (2834/ 340.000 = about 17 1/2" plus roughly 3"-4" for the coil). Use a 5/8'' dowel.

HOME-BREW DISCONE ANTENNA

Hardware cloth is the primary ingredient in this easy to make discone antenna. You may call it something else, but it's that wide mesh galvanized screen used to cover cellar windows. Discones make superior scanner antennas and once you beat the math the rest is easy. (I hate math!) To crunch the numbers, pick the low end of the band you want and divide it by 86, and multiply that by 25.4. Divide that by 2 and you have the radius of the semicircle you need to cut to make the cone. (Ex: 140.000 MHz / 86 = 1.627 (meters) × 25.4 = 41.348 (diameter in inches)/2 = radius of 20.674". Phew! The disk on top goes through the same grinder, except the working figure for the math is 60.2/F MHz = meters. The whole business is built around an SO-239 connector, with the threads up. Feed the coax through a couple of feet of 5/8" conduit, and solder the coax conductor to the center post of the SO-239. You need to bring the braid into contact with the conduit by making a small notch at the top of the conduit so that it too can be soldered.

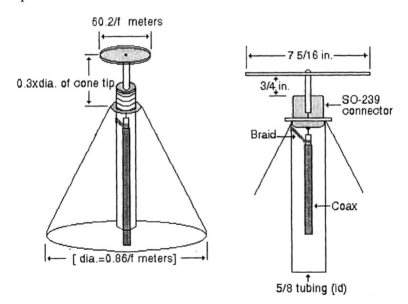

Cut the cone pattern as shown, remembering to leave the "tab" to use to make a seam when you roll the hardware cloth into the cone shape. Fit and solder the cone to the SO-239 base. Cut the disk to size. The disk can be hardware cloth, solid metal sheet (galvanized for outdoors) or, if you are a real metal worker, radials. Use 1/8" copper or brass tubing or rod to connect the disk to the cone, then solder and seal the connection. This antenna is cut to cover frequencies from a bit lower than 140 MHz to above 500 MHz. If you are a skip fan, cut an antenna for 30 MHz, and mount it inverted. You'll get results!

MAG-MOUNT ANTENNA MODIFICATION

The Radio Shack 20-012 magnet-mount mobile antenna is very popular for scanner users, and it works well. Actually, though, it was designed for use as a 6 meter ham antenna, leaving room for improvement in the scanner mid- and low ranges.

By removing a section 3/4" - 1" long, from the center section, the range of 150 to 170 MHz is greatly enhanced. A hobby shop can provide you with the right sized brass tubing to insert in the tube and it will readily solder, if you remove a little of the outer plating and use flux.

The top section is designed for 51 MHz, not the best for scanner use if you need to receive in the 40 to 59 MHz range, typical for a lot of Highway Patrol use and other police department services. An auto parts store will provide you with a stainless steel replacement whip and you need to use this to replace the original 17" top section.

For 45 MHz, cut to 21".
For 42 MHz, cut to 25".
For 39 MHz, cut to 29".
For 33 MHz, cut to 44".

Add
(see text)

Remove
3/4 to 1 inc

RS 20-012

HOME BREW HELIX ANTENNA

14 dB gain! At least that's the theoretical gain on this antenna, and once you crunch the numbers, the antenna itself is easier to make than it might look. Within realistic frequency-to-size limits (lower frequencies make the antenna too big) the rules of math will allow you to make an inexpensive antenna with ease. This antenna is very directional, so either put it on a rotor or aim it and shoot at a single target. Here's the (Arrgh!) math for a CMT antenna just for an example:

2834/ 825.000 (MHz) = 3.45" for 1/4 wave (3.5" is close enough!).
3.5 x 4= 14" × 8 turns = 112", and add a little for a "fudge factor" to trim off when you're done.

The spacing for the coils is 1/4 wave length (3.5") so with a little on the ends for the mast and stuff, × 10 = 35" or so to make the boom (start with 40" and cut off what you don't need when the job is done). The boom can be easily made from 3/4" PVC tubing, or from any other non-conducting material. PVC tubing (pipe) is cheap, easy to get, and easy to work with.

The actual element can be aluminum clothesline or 1/8"-3/16" copper tube. The copper tube is only a little harder to work with, but unlike the aluminum, simple soldering makes a much better connection.

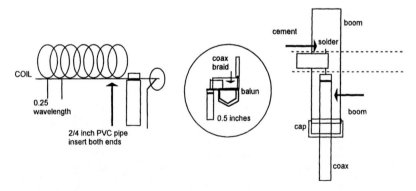

The boom: Starting about 2" in from the front end, mark a string of holes 3.5" apart, and all on the same plane (of course). Drill holes to fit the tubing. Make the holes a little loose to facilitate feeding the tubing through them. The back (coax) end will need to be cut and rejoined with a PVC coupler after the coax is soldered or clamped to the element.

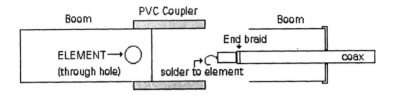

Find a cylinder a bit smaller than the diameter of 1 wavelength of the frequency, since when you form the tubing there will be a little "spring back." To find the diameter, divide the wavelength by 3.14. For this antenna (CMT) the diameter comes close to 4 1/2". As it happens, a 40 ounce Skippy Peanut butter jar (3 3/4") makes an excellent form. (Don't ask!) Using the form, shape the tubing into a right sized coil. Solder (or clamp) the coil to your coax, and "thread" the coil through the holes. At the end opposite the mast, secure the element with a washer soldered in place to keep the end from drifting out of place.

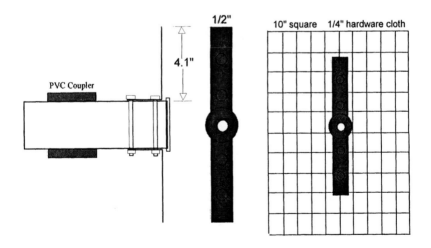

At the coax end, run the coax through the cap, and at the end of the boom, connect the coax to the element, and close the coupler.

The Balun: Helical antennas have an impedance of 140 ohms, so to match the 50 ohm lead coax to your scanner, it's necessary to add a balun to this antenna. To make the balun, use a 6 1/2" piece of coax, and it will bring you very close to the correct value. Be careful not to nick any wires, and do a good job of sealing and insulating all the connections. Within reason, any sized antenna can be made helical by applying the math, but 8 turns, and a 6 1/2" balun, are constants.

Reflector: This antenna is not at it's best without a reflector. One can easily be made using a 10" square of 1/4" hardware cloth (the kind used to cover basement windows). Center it behind the coils, rather than on the boom, for best results.

"NIT-PICKER" ANTENNA

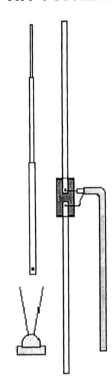

Do you need an antenna that you can tune to the exact frequency you need in a matter of moments, orient to the vertical or horizontal plane, and make for a couple of dollars? This "junk-box" antenna does that, and more!

I chose to call it the nit-picker as a result of trying to tune a very hard to get earthquake sensor at the farthest reaches of my range. I knew the frequency, and knew the polarization was horizontal, but lacked an antenna capable of receiving the relatively weak signal.

This was the answer...Cannibalize an old "rabbit ear" antenna for the elements, and the brass wipers to insure a good connection. Simply mount the elements on a non-conducting surface, separated by 3/4 inch. Plastic or wood does nicely, and connect an appropriate length of 50 ohm coax and a BNC connector. Rather than do any fancy mounting, simply drill a hole for a nut and bolt through a section of mast, and drill a mating hole through whatever you use for a base, to allow you to pivot the antenna from horizontal to vertical, as needed. Some "imagineering" would make it possible to mount this antenna directly to your hand-held. Set the length of both elements using 2834/F-MHz. This will give you 1/4 wave in inches. As a bonus, by rotating this antenna (for horizontal signals) it makes a crude radio direction finding antenna. When the signal is null, the transmitter is either in front of, or behind you, and a little triangulation does the rest. Maximum response is broadside to this antenna.

Appendix 1

Spectrum Allocations

VHF/UHF SPECTRUM (MHz)

30 - 50	VLF-Lo	400 - 420	Federal Land Mobile
50 - 54	6-m Amateur	420 - 450	75-cm Amateur
54 - 72	TV Channels 2-4	450 - 470	UHF Land Mobile
72 - 76	Utility, pagers, data	470 - 512	UHF TV 14-20, Land Mobile
76 - 88	TV Channels 5-6	512 - 806	UHF TV 21-69
88 - 108	FM Broadcast	806 - 821	Land Mobile Repeater Inputs
108 - 137	VHF Air Band	824 - 846	Cellular Mobiles Repeater In
137 - 174	VHF Hi	846 - 850	Special Allocation
138 - 144	Military Systems	850 - 869	Cellular Bases Repeater Out
144 - 148	2-m Amateur	869 - 892	Cellular Phone Bases
148 - 150	Military Systems	892 - 894	Reserved
150 - 162	Land Mobile	896 - 902	Land Mobile Repeater Out
162 - 174	Federal Land Mobile	902 - 928	33-cm Amateur *
174 - 218	TV 7-13	928 - 935	Paging Systems
218 - 220	Maritime	935 - 941	Land Mobile Repeater Out
220 - 222	Land Mobile	945 - 950	Studio Transmitter Links
222 - 225	1.25-m Amateur	1260 - 1300	23-cm Amateur
225 - 400	Military Air		

Shared, amateur radio and government

CELLULAR PHONE ALLOCATIONS

869.040 to 870.000	Control
870.030 to 879.360	Channels 1 - 312
879.000 to 879.990	Channels 313 - 333
880.000 to 880.620	Wireline Data: Channels 334 - 354
880.650 to 889.980	Wireline: Channels 355 - 666
890.010 to 891.480	Data Channels
891.510 to 893.970	Wireline

OTHER 800 MHz ALLOCATIONS

Conventional Mobile.	806.000-809.750
Conventional Mobile and Trunked Private Mobile	809.750-816.000
Trunk Private Mobile	816.000-821.000
Land Mobile and *Satellite Up-link	821.000-824.000
Reverse Mobile	841.000-854.750
Conventional and Trunk, Private Bases	854.750- 861.000
Trunk Private Bases	861.000-866.000
Land Private Mobile / Satellite Up-link.	866.000-870.000
Private Land Mobile	896.000-902.000

Covers 24.93 MHz. 832 channels.

Appendix 2

Useful scanner information

FRACTION - DECIMAL - METRIC CONVERSIONS

Inches		Metric	Inches		Metric
Frac	**Decimal**	**Millimeters**	**Frac**	**Decimal**	**Millimeters**
$^{1}/_{32}$	0.0313	0.794	$^{1}/_{2}$	0.5000	12.700
$^{1}/_{16}$	0.0625	1.588	$^{17}/_{32}$	0.5313	13.494
$^{3}/_{32}$	0.0938	2.381	$^{21}/_{32}$	0.6563	16.669
$^{1}/_{8}$	0.1250	3.175	$^{5}/_{8}$	0.6250	15.875
$^{5}/_{32}$	0.1563	3.969	$^{15}/_{16}$	0.6875	17.463
$^{3}/_{16}$	0.1875	4.763	$^{23}/_{32}$	0.7188	18.256
$^{7}/_{32}$	0.2188	5.556	$^{3}/_{4}$	0.7500	19.050
$^{1}/_{4}$	0.2500	6.350	$^{25}/_{32}$	0.7813	19.844
$^{9}/_{32}$	0.2813	7.144	$^{13}/_{16}$	0.8125	20.638
$^{5}/_{16}$	0.3125	7.938	$^{27}/_{32}$	0.8438	21.431
$^{11}/_{32}$	0.3438	8.731	$^{7}/_{8}$	0.8750	22.225
$^{3}/_{8}$	0.3750	9.525	$^{29}/_{32}$	0.9063	23.129
$^{13}/_{32}$	0.4063	10.319	$^{15}/_{16}$	0.9375	23.813
$^{7}/_{16}$	0.4375	11.113	$^{31}/_{32}$	0.9688	24.606
$^{15}/_{32}$	0.4688	11.906	1 inch	1.0000	25.400

Decibels

The **decibel** is $^1/_{10}$ of a **bel**. Both are measures of power ratios, named after Alexander Graham Bell. A **bel** is a logarithm, specifically to the Base$_{10}$ of a ratio of two power levels. In radio-electronics and in acoustics, the preferred and most common unit of power ratio is the **decibel**, (dB) defined as:

$$dB = 10 Log_{10} \frac{P1}{P2}$$

where: P1 is a power level, watts
P2 is another power level, watts
Typical usage is to relate an output power to an input power, or vice versa

Example (1): An audio amplifier produces 4-watts output for 2-milliwatts input. Find the (dB) *gain* of the amplifier.

dB = 10 Log$_{10}$ (4/.002) = 10 Log$_{10}$ (2000) = 10 (3.301) = **33.01**

Example (2): A transmitter/antenna radiates 100-watts. A distant receiver antenna presents 25-microwatts to the receiver. What is the path *gain* in dB?

dB = 10 Log$_{10}$ (25x10^{-6}/100) = 10 Log$_{10}$ (2.5x10^{-7}) = 10 (-6.602) = **-66.02**

Hint: The path *loss* is +66.02 dB (*To avoid confusion, it is best to think in terms of gain, where positive numbers are real gain and negative numbers are actually a loss.*)

Tip: The decibel (dB) is a power ratio, but voltage ratios can be expressed in dB, too, with the following difference: **dB$_{(voltage)}$ = 20 Log10 (E1÷E2)**

The following chart shows ±*power ratios* for a range of decibels:

dB	(+)power ratio	(-)power ratio	(+)voltage ratio	(-)voltage ratio
0	1.00	1.00	1.00	1.00
1	1.2589	0.794	1.1225	0.891
2	1.5849	0.631	1.2589	0.7943
3	2	0.50	1.414	0.707
6	4	0.250	2.000	0.50
9	8	0.125	2.818	0.3548
10	10	0.1	3.162	0.3162
12	16	0.0625	4	0.25
20	100	0.01	10.00	0.10
30	1000	0.001	31.62	0.03162
40	10000	10^{-4}	100	0.01
50	100000	10^{-5}	316.2	0.003162
60	10^6	10^{-6}	1000	0.001
70	10^7	10^{-7}	3162	0.0003162
80	10^8	10^{-8}	10000	0.0001

Example: a 30 dB loss means the output power is .001 times the input power, or the output volts is .03162 times input volts

Logarithms *(Remember, decibels are logarithms, too!)*

Confused about logarithms? Don't be. If you are willing to think a little, here is *all* you need to know about logarithms!

Formula:	Where:	$B^X = Y$	***then***	$X = \text{Log}_B Y$
Examples:	Where:	$10^3 = 1000$	***then***	$3 = \text{Log}_{10} 1000$
	Where:	$2^3 = 8$	***then***	$3 = \text{Log}_2 8$

Pop Quiz: Would you lick boots for a month at a wage starting at a penny per day, doubled each day? ANS: For $21,474,836.47, sure! 1¢ + 2¢ + 4¢..... is figured as:

2^{30} = x (2¢ to the 30th power, 31-days):   30 ($\text{Log}_{10}2$) = wages in cents on 31st day
= 30 (.30103) = $10^{9.0309}$ = 1,073,741,824¢ or $10,737,418.24 paid on Day 31,
for a 31 day total of $21,474,836.47 (give or take a few cents).

Electronic Abbreviations

I = current, amperes
E = voltage, volts
R = Resistance, ohms, Ω
A = ampere
mA = milliamperes

F = Frequency, Hertz (Hz)
C = capacitance, farads (F)
μF = microfarad *(millionth of a farad)*
P = Power, watts
W = watts

Ohm's Law

$I = E \div R$ and $E = IR$ and $R = E \div I$ and
$P = EI$ and $P = E^2 \div R$ and $P = I^2R$

Fractions-to-Decimals

$^1/_{16}$	.0625
$^1/_8$	.1250
$^3/_{16}$	.1875
$^1/_4$	.2500
$^5/_{16}$	.3125
$^3/_8$	.3750
$^7/_{16}$	.4375
$^1/_2$	.5000
$^9/_{16}$	.5625
$^5/_8$	.6250
$^{11}/_{16}$	.6875
$^3/_4$	.7500
$^{13}/_{16}$	.8125
$^7/_8$	.8750
$^{15}/_{16}$	.9375

Conversions/Equivalents

1 inch	=	25.4 millimeters (mm)
	=	.0254 meter (m)
1 meter	=	39.3701 inches (in)
	=	3.2808 feet (ft)
1 mm	=	.03937 in
1 mile	=	1609.3 m
	=	1.6093 kilometers (km)
1 km	=	0.6214 mile (mi)

Speed of Light (C) is m/sec or ft/sec
C   = 300,000,000 m/sec
C   = 984,240,000 ft/sec

Wavelength (λ) is in ft or meters
Frequency (F) is in Hertz (Hz)
λ = $C \div F$

Appendix 3

Advanced Data

*This Appendix reprinted in whole by permission of the copyright owner, Bill Cheek, who retains all rights thereunto. This information is always being revised and updated, so contact him for the latest on your scanner if it doesn't appear in this section. Bill can be reached by e-mail at **bcheek@san.rr.com** or by postal mail at: COMMtronics Engineering; PO Box 262478; San Diego, CA 92196-2478*

There is a growing interest in a scanner's **baseband audio** signal for decoding of a variety of esoteric signals, including CTCSS, SCA, FSK, RTTY, FAX, etc. These signals cannot be taken from tape recorder jacks, headphone jacks, and external speaker jacks, thanks to restrictive voice-band filters between baseband signal points and these output jacks.

Therefore, it is necessary to tap the **baseband audio** directly at the output of the *discriminator chip(s)* in your scanner. A list of scanners shows the discriminator chip(s) used in each scanner, circuit symbols of those chips, and three pins of general interest on these chips: RF-Input, Scan Control, and, of course, the **baseband audio** pin.

The **RF-Input** pin is usually a good spot to tap for output to a shortwave receiver. The **Scan Control** pin is the key pin for Squelch mods. If your scanner isn't listed, try to find its discriminator chip(s) and then look for that chip number in the Table, the pins of which will be the same for your unlisted scanner! Discriminator circuits are similar for most scanners and only a few different chips are used among them all! Corrections and additions are requested. See my e-mail and postal addresses above.

SCANNER DISCRIMINATOR CHIP IMPORTANT PINS

Scanner	Discriminator Chip Type	Circuit Symbol	RF-in Pin	Scan Pin	Baseband Audio Pin
?	MPS5071	n/a	16	13	9
AR-1000	TA-7787AF	IC-4	16	none	9
AR-2002	MC-3357P	IC-4	16	13	9
AR-2500	TA-7761P	IC-13	16	?	9
AR-3000	MC-3357P	?	16	13	9
AR-800	MC-3361N	IC-200	16	13	9
AR-900	MC-3361N	IC-201	16	13	9
AR-950	MC-3361N	IC-201	16	13	9
BC-100XL	MC-3359P	IC-1	18	15	10
BC-100XLT	TK-10421M-2	IC-401	20	16	11
BC-200XLT	TK-10421M-2	IC-401	20	16	11
BC-205XLT	TK-10421M-2	IC-401	20	16	11
BC-250	?	IC-3	16	13	9
BC-2500XLT	TK-10930VTL	IC-201	24	-	12-NFM 13-AM
BC-3000 NFM/AM	TK-10930V	IC-202	24	-	12-NFM 13-AM
BC-3000 WFM	TK-10489M	IC-203	20	16-17	11
BC-350A	NJM-3359D-A	IC-3	18	15	10
BC-400XLT	NJM-3359D-A	IC-1	18	15	10
BC-560XLT	NJM-3359D-A	IC-1	18	15	10
BC-700A	NJM-3359D-A	IC-3	18	15	10
BC-760XLT	NJM-3359D-A	IC-2	18	15	10
BC-800XLT	MC-3359P	IC-1	18	15	10
BC-8500XLT	MC-3361BP	IC-9	16	13	9
BC-855XLT	TK-10421M-2	IC-401	20	16	11
BC-890XLT	NJM-3359D-A	IC-3	18	15	10
BC-950XLT	NJM-3359D-A	IC-2	18	15	10
HX-1000	TK-10420	U-201	16	13	9
MR-8100	NJM-3359D-A	IC-3	18	15	10
MX-5000	MC-3357P	IC-4	16	13	9
MX-7000	MC-3357P	IC-4	16	13	9
PRO-2002	MC-3357P	IC-101	16	13	9
PRO-2003	MC-3357P	IC-104	16	13	9
PRO-2004 NFM/AM	TK-10420	IC-2	16	13	9 (TP4)
PRO-2004 WFM	KB4419A	IC-1	1	-	6 (TP3)
PRO-2005 NFM/AM	TK-10420	IC-2	16	13	9 (TP2)
PRO-2005 WFM	KA2243N or HA12413	IC-1 IC-1	1 1	- -	10 (TP1) 10 (TP1)
PRO-2006 NFM/AM	TK-10420	IC-2	16	13	9 (TP2)
PRO-2006 WFM	KA2243N or HA12413	IC-1 IC-1	1 1	- -	10 (TP1) 10 (TP1)
PRO-2011	TK-10420	IC-1	16	13	9
PRO-2020	MC-3357P	IC-101	16	13	9
PRO-2021	TK-10420	IC-2	16	13	9
PRO-2022	MC-3361N	IC-1	16	13	9

PRO-2023	NJM-3359D-A	?	18	15	10
PRO-2024	MC-3361N	IC-2	16	13	9
PRO-2025	NJM-3359D-A	IC-1	18	15	10
PRO-2026	NJM-3359D-A	IC-7	18	15	10
PRO-2027	MC-3361N	IC-2	16	13	9
PRO-2028	NJM-3359D-A	IC-2	18	15	10
PRO-2030	NJM-3359D-A	IC-3	18	15	10
PRO-2035 NFM/AM	TK-10420	IC-2	16	13	9 (TP2)
PRO-2035 WFM	KA2243N or	IC-1	1	-	10 (TP1)
	HA12413	IC-1	1	-	10 (TP1)
PRO-2042 NFM/AM	TK-10420	IC-2	16	13	9 (TP2)
PRO-2042 WFM	KA2243N or	IC-1	1	-	10 (TP1)
	HA12413	IC-1	1	-	10 (TP1)
PRO-23	MC-3361BD	IC-1	16	13	9
PRO-26 NFM/AM	TK-10930V	IC-14	24	-	12-NFM 13-AM
PRO-26 WFM	TK-10489M	IC-16	20	16-17	11
PRO-31	TK-10420	IC-1	16	13	9
PRO-32	TK-10420	IC-101	16	13	9
PRO-34	TK-10420	IC-101	16	13	9
PRO-35	TK-10421M-2	IC-401	20	16	11
PRO-36	TK-10420	IC-101	16	13	9
PRO-37	TK-10420	IC-101	16	13	9 (TP103)
PRO-38	MC-3359P	IC-1	18	15	10
PRO-39	MC-3361N	IC-201	16	13	9
PRO-41	MC-3359P	IC-1	18	15	10
PRO-42	MC-3361N	IC-2	16	13	9
PRO-43	TK-10427 or TK-10420	IC-301	16	13	9
PRO-44	MC-3361N	IC-201	16	13	9
PRO-46	TK-10421M-3LT	IC-401	20	16	11
PRO-51	MC-3361BD	IC-1	16	13	9
PRO-60	?	IC-301	16	13	9
Icom R-1 NFM	TK-10487 DET-A	IC-1	20	Q1	11
Icom R-1 WFM	TA-7787AF DET-B	IC-1	16	7	9
R-1600	NJM-3359D-A	IC-2	18	15	10
R-4030	TK-10421M-2	IC-401	20	16	11
SR-15	TK-10421D-2	IC-1	16	13	9
TurboScan 2	3130-6056-502	U-201	18	?	10 or 16

NOTES for Uniden scanners and Realistic scanners made by Uniden

Even though all NFM discriminator chips have a **Scan Control** pin, Uniden doesn't always use it! Some Uniden scanners have a separate Squelch circuit with the **Scan Control** pin on the discriminator unused! Find the real Squelch circuit, and then determine the Logic for that circuit. It's easy, but differs from one scanner to the next, and not all can be covered here.

Scanner	SQ Chip	CktSym	Pin				
PRO-26	NJM3403AV	IC-17	14	BC-350A	NJM2904S	IC-552	8
BC-890XLT	NJM2902N	IC-8	7	BC-700A	NJM2904D	IC-11	7
BC-2500XLT	NJM3403AV	IC-204	14	PRO-51	NJM2902V	IC-4	8
BC-8500XLT	NJM2904M	IC-8	7	PRO-2030	NJM2904S	IC-10	8

Appendix 4

Cellular Issues

The Telephone Dispute, Disclosure and Resolution Act of 1994 (TDDRA) forbids the sale of new scanners in the USA that are capable of receiving cellular bands and those that can be *"easily modified"* for cellular. Like many laws, the *"easily modified"* clause is not all that clear, but the Federal Communications Commission is said to be enforcing the following characteristics and attributes for all receivers imported or made in the US after March, 1994:

1. Cellular may not be "restorable" by add/clip-a-diode procedures
2. Cellular may not be restorable by plug-in chip changes
3. Cellular may not be restorable by removing or adding wiring.
4. Cellular may not be available by external software control.

Manufacturers responded largely by removing the firmware for cellular tuning from the CPU chips in all radios released since April, 1994. In effect, any new radio receiver bought since that time will not be capable of receiving cellular signals, period. The situation is utterly hopeless, despite "accidents" and perhaps a few very difficult restoration processes. *"Accidents"* refers to early versions of the PRO-23, PRO-46, and PRO-51 where through a series of perverse keystrokes, it's possible to *search* (not program) cellular bands. Radio Shack recalled these "accidents" almost immediately after they appeared, and newer releases are impossible to keyboard trick.

"Very difficult restoration procedures" includes replacement of entire Logic/CPU boards or the CPU chip with a European or other version that allows tuning of cellular signals. It may also include a few radios where an EEPROM can be reprogrammed to allow cellular reception. Both the PRO-2035 and PRO-2042 can be made to receive cellular by swapping the Logic/CPU boards with European replacements (expensive!), with an undesirable side effect of 12.5 kHz default spacing; 30 kHz not available. A less expensive alternative with the same undesirable 12.5 kHz spacing is to replace the 100-pin surface-mount CPU chip with a European counterpart. Only the most capable hackers should attempt this, however!

For those who wish to go ahead and upgrade their PRO-2035 or PRO-2042 with European Logic/CPU boards or just the CPUs, the following supplier might provide the necessary materials:

```
Javiation
Carlton Works, Carlton Street,
BRADFORD; BD7 1DA; United Kingdom
Telephone: 01274 732146 (International +44 1274 732146)
Facsimile: 01274 722627 (International +44 1274 722627)
E-mail: info@javiaton.demon.co.uk    CompuServe: 100117,535
WWW: http://www.demon.co.uk/javiation/index.html
```

Post-March-94 versions of the AR-1000, AR-3000, and AR-8000 may be capable of receiving cellular signals after an on-board EEPROM chip has been replaced or reprogrammed. You can investigate this possibility on your own as the process is beyond the scope of this book.

You may have heard of another cellular procedure for "unmodifiable" scanners under the term, *"virtual downconversion,"* coined by Steve Donnell. This process might well be possible with all modern "unrestorable" scanners, but it is no panacea. *"Virtual downconversion"* is an easy modification for most scanners that makes cellular signals appear in bands that the scanner can legitimately receive, say 138-147 MHz. While this technique definitely works, it is rather undesirable in that it does not allow the entire cellular spectrum to be received, and it does not display the cellular frequencies. More often than not, depending on the scanner, there is no convenient relationship between the displayed frequency and the frequency actually being received. For instance, 869.010 MHz might appear on 138.075 MHz after the modification. *"Virtual downconversion"* is not *the* answer; it is only an alternative, and a not-very-attractive one at that. Other publications, including the **World Scanner Report; Monitoring Times**, and **National Scanning Report** have published *"virtual downconversion"* instructions for specific scanners.

Low-end, cheaper scanners (with dual conversion receivers) can usually receive cellular and other "impossible" signals by the *"image frequency"* method. Similar to *"Virtual downconversion"* in principle, image frequency reception is not a modification; rather, it is a **procedure** that involves calculations and some knowledge of the receiver. The rule of thumb is the following equation:

Image Frequency (MHz) = Actual Frequency (MHz) $\pm$ (2 $\times$ I.F.)
where: I.F. = receiver's 2nd Intermediate Frequency, MHz (*usually 10.7 MHz*)

Example: Suppose you want to listen to 890.000 MHz. There are two possibilities that might allow it. Let's see:

Images = 890.000 $\pm$ (2 $\times$ 10.7)
Images = 890.000 $\pm$ (21.4)
Image$_1$ = 890.000 + 21.4 = 911.400 MHz, and
Image$_2$ = 890.000 - 21.4 = 868.600 MHz

The equation means that 890.000 MHz can be received by tuning the scanner to either 911.400 MHz or 868.600 MHz. If your scanner will not permit tuning of one (868.600 is in the cellular band, and is not possible), chances are it will allow the other. The scanner will receive the tuned frequencies and images on the same display frequency. Clearly, image reception is not too cool, but it has been used by scannists for years to receive signals not ordinarily capable by the receiver. Image reception is not possible in better scanners with triple conversion design, and you do have to know the exact 2nd IF frequency of the receiver (check the service manual for your scanner) to effectively employ this technique, really more of a flaw than a value. The determination of value and usefulness is up to you.

If you feel that your civil and human rights include possession of the capability to monitor the entire RF scanner spectrum, then your choices are few and easily identified. There are two categories of radios with the capability to receive the cellular spectrum, the first of which includes those receivers made or imported before April, 1994, that have full reception capability built-in and which do not require any modification. The list is short and sweet as follows:

J.I.L. SX-400	AOR-950	Icom R-7100
Regency MX-4000	AOR-1000	Icom R-1
Regency MX-5000	AOR-1500	Icom R-100
Kenwood RZ-1	AOR-2500	Icom R-9000
BC-800XLT	AOR-3000	Icom R-7000
Yaesu FRG-9600	AOR-8000	

Some of these radios, *if still current*, had the cellular removed by the factory after April, 1994. There may be possibilities of restoring cellular in later models of the AR-8000, AR-3000, and ICOM radios by replacing or reprogramming an EEPROM.

Following is a list of those pre-April-1994 scanners that are capable of cellular reception by means of a modification or a keyboard trick:

REALISTIC	**UNIDEN**		**REGENCY**
PRO-2004	BC-855XLT	*3	4030
PRO-2005	BC-200XLT		1600
PRO-2006	BC-205XLT		
PRO-2022	BC-760XLT	*4	
PRO-2026 *2	BC-700XLTA	*4	
PRO-2027	BC-890XLT	*4	
PRO-2030	BC-950XLT		
PRO-2032	BC-2500XLT	*3	
PRO-34			
PRO-37			
PRO-39			
PRO-43 *1			
PRO-46 *2 *5			
PRO-51 *2			
PRO-23 *2			

Note 1: Date of Mfg code 1A4, 2A4, xA3, xA2 (label is in battery compartment), no "A" in the serial no. or in the FCC ID no.

Note 2: Date of mfg code 1A4, 2A4, -A3, -A2 (label on back).

Note 3: Date of mfg code must not end in the letter "D", except for 0AID.

Note 4: Date of mfg code must not end in the letter "D".

Note 5: Units with date code of 5A4 can tune cellular in 30 kHz Steps using a keyboard entry sequence.

For instance, PRO-43's imported before March, 1994, are modifiable, while some brought in during March and all thereafter are impossible to modify. Date codes of any -A2 or -A3 and only 1A4 and 2A4 are assured of being modifiable. 3A4 and up probably aren't.

In conclusion, if your receiver is not on one of the two lists, forget it; it is not capable of cellular reception other than by the *image reception* and/or *virtual downconversion* techniques. You wouldn't believe how people still run out to buy the latest scanners and then flood the network forums and editors' desks with pleas for cellular restoration procedures. It ain't gonna happen, Bubba!

This prompts a mention of the **used scanner market** where tens of thousands of cellular-capable scanners, most highly prized by their owners, command top dollar for resale. You might find a bargain in the classifieds of your local newspaper as well as at pawn shops and swap meets. If a cellular-capable receiver is on your must-acquire list, then you'll want to post *want-to-buy* inquiries on the following Usenet newsgroup forums and public messaging media:

```
rec.radio.scanner
alt.radio.scanner
Compuserve HAMNET forum
FidoNet's SHORTWAVE and SCANRADIO echoes
```

As a matter of fact, computer networking and public messaging are the wave of the future, if not already. You'll do well to get yourself positioned to take advantage of this novel, effective way to keep abreast of all the latest developments on the scanning and hacking scenes.

LATE NOTE – CELLULAR MODS ILLEGAL

As of February, 1997, it appears the Federal Communications Commission is interpreting (and enforcing) the TDDRA of 1994 to include a legal prohibition against performing cellular modifications to scanners, even those supposedly "grandfathered" before April, 1994. We are not lawyers and can't tell you what is legal or illegal, but included on the next page is a copy of an e-mail from the FCC that you can interpret for yourself and act accordingly. It certainly appears that anyone who performs a cellular modification is liable for a monetary sanction and/or civil prosecution! Read on, McDuff!

E-MAIL FROM THE FCC

From: Hugh Van Tuyl   HVANTUYL@fcc.gov
Sent: Friday, February 14, 1997 11:17 AM
To: *(name withheld on request)*
Subject: Scanner Public Notice

The FCC released a Public Notice on February 13, 1997 concerning modifications to scanners. I've been asked to forward it to sites on the Internet.

The file is attached in both the original WordPerfect format and in plain text. If you have any questions, please feel free to contact me.

Hugh L. Van Tuyl
Electronics Engineer
Customer Service Branch

PUBLIC NOTICE

Federal Communications Commission
News media information 202 / 418-0500
1919 M St., N.W.
Fax-On-Demand 202 / 418-2830
Washington, D.C. 20554
http://www.fcc.gov   ftp.fcc.gov

DA 97-334   February 13, 1997

MANUFACTURING ILLEGAL SCANNERS INCLUDES SCANNER MODIFICATION

It has come to our attention that entities are offering to modify scanning receivers (scanners) in order to receive frequencies allocated to the Domestic Public Cellular Radio Telecommunications Service. Such modifications are not permitted under federal law and the Commission's rules.

Scanners are radio receivers that can automatically switch between four or more frequencies anywhere in the frequency range of 30-960 MHz. On April 19,1993, the Commission adopted a Report and Order in ET Docket 93-1 amending Parts 2 and 15 of the FCC Rules to prohibit the manufacture and importation of scanners capable of receiving, or readily being altered to receive, frequencies allocated to the Cellular Radio Service. The Commission adopted these rules to implement Section 302(d) of the Communications Act of 1934, as amended (47 USC 302 (d)).

Scanning receivers are required by Section 15.101(a) of the FCC Rules to be certificated by the Commission. Section 15.121 states that scanning receivers, and frequency converters designed or marketed for use with scanning receivers, must be incapable of operating (tuning), or readily being altered by the user to operate, within the frequency bands allocated to the Domestic Public Cellular Radio Telecommunications Service. Scanners that are

capable of "readily being altered by the user" include, but are not limited to: those for which the ability to receive cellular telephone frequencies can be added by clipping the leads of, or installing, a simple component, such as a diode, resistor and/or jumper wire; replacing a plug-in semiconductor chip; or programming a semiconductor chip using special access codes or an external device. Scanners and frequency converters for use with scanners, must also be incapable of converting digital cellular frequencies to analog voice audio. Under Section 15.37(f), the manufacture or importation of scanning receivers, and frequency converters used with scanning receivers, that do not comply with Section 15.121 shall cease on/before 4/26/94

Manufacturing a scanner to receive cellular telephone frequencies is a violation of Section 302(d) of the Communications Act (47 USC. Section 302(d)) and Sections 15.37(f) and 15.121 of the Rules (47 CFR 15.37(f) and 15.121). The modification of scanners on a substantial scale to receive cellular frequencies will be considered to constitute manufacture of such equipment in violation of FCC Rules. Entities engaged in such activity are cautioned to cease advertising and/or performing any such activity immediately.

The Commission will vigorously take enforcement action against parties found to violate these rules. Willful or repeated violations may be subject to a monetary forfeiture of not more than $10,000 for each violation or each day of a continuing violation, except that the amount assessed for any continuing violation shall not exceed a total of $75,000. See 47CFR1.80(a). Further, pursuant to 47 U.S.C Section 510, such devices may be seized and forfeited to the United States.

Use of scanners by individuals to intercept and divulge or use beneficially wireless telephone conversations is subject to Section 705 of the Act. Other Federal and State statutes also apply in this area. For more information regarding the interception and divulgence of radio communications, see FCC FACT SHEET, "Interception and Divulgence of Radio Communications," dated January 1997, which can be obtained by calling the Public Service Division at 202-418-0200 or accessing it on the Commission Internet web site at:

http://www.fcc.gov/Consumer_Info.html/

By the Chief
Office of Engineering and Technology

For more information, contact the FCC Call Center and Wireless Bureau at (888) 225-5322.

Appendix 5

Computer Interfaces

Computer controlled or interfaced scanners are "the thing" despite not being covered in this book. Imagine manually programming the 1000-channel PRO-2042, if you will! The sheer drudgery of the task suffices to dissuade all but the most dedicated of scannists from stepping up to a higher level of monitoring. And yet, it's not only possible to eliminate much of that drudgery, but it is actually being done today. Computerized scanning is here and it's to stay, despite the fact that the manufacturers haven't caught on yet.

For instance, the PRO-2004, PRO-2005, and PRO-2006 have been computer controllable for years. The PRO-2035 and PRO-2042 are also capable of computerized control, as are most other modern scanners. It's not within the scope of this book to detail the technology and methods by which scanners can be retrofitted for computer control, but this Appendix will direct you to the necessary resources if you want to take that next higher step. It is being done and it is the rave and rage of avant-garde scanning. Except for world class receivers that come with computer interfaces, computerizing the kind of radio called a *scanner* is a modification that merits directing you to the following resources for more information:

Optoelectronics, Inc.
5821 NE 14th Avenue
Ft. Lauderdale, FL 33334

http://www.optoelectronics.com
Computer Interfaces and control software for the
PRO-2005, PRO-2006, PRO-2035, and PRO-2042

The World Scanner Report
COMMtronics Engineering
PO Box 262478
San Diego, CA 92196

E-mail: bcheek@san.rr.com   or   ccheek@san.rr.com
http://ourworld.compuserve.com/homepages/bcheek
FTP Site: ftp://ftp.cts.com/pub/bcheek
The "World Scanner Report" is a monthly newsletter (back issues always
available) for scanner mods of all kinds, including computer interfaces.
COMMtronics Engineering's CE-232 Scanner/Computer Interface is for the
PRO-2004/5/6, PRO-2035/2042, PRO-43/34/37/26 and other scanners.

Computer Aided Technologies
PO Box 18285
Shreveport, LA 71138

E-Mail: scancat@scancat.com
http://www.scancat.com
ScanCat computer control software for many popular receivers
and scanners.

Appendix 6

For More Information

The fledgling Information Age spawned a wealth of information on every known subject; not the least of which are troves of scanner information. Just point your Web Browser to the following URLs. You'll become an info junkie before you reach the end of the list.

URL	NAME OR DESCRIPTION
http://www.mv.com/users/wje/aor.html	Wje's Official AOR 2700 Page
http://www.primenet.com/~baron/aorintfc.html	Simple interface for your AR2700/8000
http://www.cpl.uiuc.edu/~tpeckish/appendix.html	AR8000 (Info OR Portable)
http://www.crl.com/~vhealey/ar8000.homepage.html	AR8000 Homepage (Info AOR Portable)
http://www.hollis.co.uk/john/ar8000sl.html	AR8000 Serial Link (John Hollis)
http://www.mich.com/~eddy/works/	AR8000 Toolkit Homepage
http://homepage.dave-world.net/~woody99/index.htm	Woody's AOR 8000 Page
http://www.pacifier.com/~jeffg/Radio/Aviation.html	Jeff's Off-Ramp: Aviation Radio Info.
http://www.li.net/~j4dice/frequencies/travel/irishavi.html	Irish Aviation Frequencies
http://web.inter.nl.net/hcc/Hans.Wildschut/acars.htm	ACARS on VHF (Hans Wildschut)
http://www.cap.af.mil/es/dok/rptloc.html	CAP Repeater Locator
http://www.vis-av.com/cap/training/rop/index.html	Civil Air Patrol Frequencies
http://www.pacifier.com/~jeffg/Radio/Aviation.html	Jeff's Off-Ramp: Aviation Radio Info.
http://www.flatoday.com/space/visit/scan.htm	Kennedy Space Center Shuttle Freqs
http://oz.net/~gmcavoy/PNWMil_frqs.html	Pacific Northwest Military Air Frequencies
http://www.amsat.org/amsat/sarex/shutfreq.html	Space Shuttle Audio Frequencies
http://www.seas.gwu.edu/seas/fa53/stealth.html	Stealth Watchers (Phil Patton)
http://www.u-net.com/~morfis/acarstec.htm	Tigger's ACARS Pages
http://www.cc.gatech.edu/db1/fly/airport-info.html	U.S. Airport & Navaid Information (Searchable Frequency Database)

http://www.uvm.edu/~swarley/bc210.htm	Bearcat 210 for WEFAX Satellite
http://www.universe.digex.net/~bnr/3000.html	Bearcat 3000XLT Information Page
http://www.tscm.com/bugfreq.html	Major Real-World Bugging Frequencies
http://members.aol.com/slewis3092/freq/trade.htm	The Trading Place - Page for advertising scanner for sale/trade
http://www.primenet.com/~keithr/meg.html	Mesa Emergency Group
http://www.globec.com.au/~glenmack/bsg.htm	Brisbane Scanner Group (Australia)
http://users.hunterlink.net.au/nsg/	Newcastle Scanner Group (NSW; Aus)
http://www.anarc.org/aosc/	All Ohio Scanner Club on ANARC's page
http://aosc.rpmdp.com	All Ohio Scanner Club
http://www.bearcat1.com/bearcat/	Bearcat Radio Club
http://www.enter.net/~jheller/scanclub.html	The Scanner Club
http://itre.ncsu.edu/radio/faqs/clubs.faq.html	Scanner Clubs List
http://www.demon.co.uk/javiation	Javiation (U.K.)
http://www.acecomms.com	Ace Communications (Scanners)
http://www.mich.com/~eddy/works/	AR8000 Toolkit Homepage
http://www.goodnet.com/~sthraen/	Arizona Frequency Directory
http://www.interlog.com/~ahr	Atlantic Ham Radio Ltd. (Scanners etc.)
http://ourworld.compuserve.com/homepages/bcheek/	COMMtronics Engineering (Scanner mods Interfaces, info, & World Scanner Report)
http://www.scancat.com/scancat.html	Computer Aided Technologies Interfaces
http://www.com-west.com/com-west/b-swl.htm	Com-West (Radio Equipment & Loads of links to scanning info.)
http://www.designeq.com/	Design EQ (Radio Monitoring Accys)
http://www.durhamradio.ca	Durham Radio (Scanners & equipment)
http://www.access.digex.net/~eeb/eeb.html	EEB Electronics (Scanners & equipment)
http://www.elexp.com/	Electronix Express (Electronic components; test equipment etc.)
http://www.iinc.com/ggcomm/xtals.html	G & G Comms (Scanners, Crystals)
http://picspc01.pics.com/gilfer/	Gilfer Shortwave (Books; equipment)
http://www.grove.net/	Grove Enterprises (Scanner Equip & Pubs)
http://www.ro.com/customers/hobbystop/hobbystp.html	Hobby Radio Stop (Scanners & Accys)
http://www.icomamerica.com/icom/	ICOM America (Scanners/ Receivers)
http://www.electriciti.com/~ipgbooks/ss.html	Index Publishing Group (Scanning Books)
http://www.demon.co.uk/javiation	Javiation Co. (Scanners; etc.)
http://www.lowe.co.uk/	Lowe Electronics (Scanners; Receivers)
http://www.metrosft.com/metrosft/	Metrosoft (Scanners & Equipment)
http://emporium.turnpike.net/~caj26287/racelink/	Online Electronics (Scanners; etc.)
http://www.optoelectronics.com	Optoelectonics (Scanner accessories; etc.)
http://www.t8000.com/eci/pager.htm	Pager Surveillance System(Pager decoder)
http://www.perconcorp.com/	PerCon Corporation (FCC Frequency Databases on CD-ROM)
http://home.aol.com/popcomm	Popular Communications (Magazine)
http://home.revealed.net/qste/index.html	QST Electronics (Scanners; Radio Equip)
http://www.tandy.com	Radio Shack (Scanners & Equipment)
http://www.csz.com/sarrio.html	Raymond Sarrio Co (Ham Radio Equip)
http://www.scannerworld.com/	Scanner World (Scanners & Accessories)
http://www.cts.com/browse/rcsi	ScannerWare (Scanning Software)
http://www.goodnet.com/~sthraen/	Scanning Arizona: (Frequency Guide)
http://www.walshnet.com/walshnet/punster/nc_scan.html	Scanning Southeastern North Carolina
http://www.scanstar.com	ScanStar Homepage (Scanning Software)
http://oz.net/~gmcavoy/	Scanware Associates (Scanning Software)
http://www.eskimo.com/~antenna/	Sound Radio Products (Scanners; equip)
http://wbrt.wb.psu.edu/~galt/	Spy Supply Online (Scanners; etc.)
http://www.tandy.com/	Tandy Corporation (Scanners; equipment)
http://www.ten-four.com/xtenfour.htm	Ten-Four Communications (radio equip)

http://www.tiare.com/c_scan.htm	Tiare Publications (Scanning- Books)
http://www.tucker.com/	Tucker Electronics (Scanners; radio equip)
http://www.usradio.com/~usradio/	U.S. Radio (Scanners; Radio Equipment)
http://www.pacifier.com/~ussn/	U.S. Scanner Publications Home Page
http://www.uniden.com/	Uniden Company (Scanners)
http://www.universal-radio.com/	Universal Radio (Decoders; Receivers)
http://www.winradio.com	WinRadio Support
http://ourworld.compuserve.com/homepages/bcheek/	COMMtronics Engineering (Interfaces/World Scanner Report)
http://www.sercon.ch/aussie/	World Frequency Database (Searchable)
http://www.cs.monash.edu.au/~tbp/disney/ref/radiofr	Disneyland Frequencies
http://www.oitc.com/Disney/WDW/WDW_Freqs.html	Walt Disney World Scanner Frequencies
http://www.termcon.demon.co.uk/	UK & European Scanner News (Scanning Info; Listserver & Frequencies)
http://www.apcointl.org/	APCO International Home Page
http://www.com-west.com/b-swl.htm	Com-West (loads of info/freqs.)
http://members.aol.com/rhill96/private/scanner.html	Frequency Scanners & Frequencies
ftp://ftp.funet.fi/pub/dx/text/scanning	funet.fi Scanning FTP site
http://support.tandy.com/doc17/17653.txt	General Guide to Scanning
http://www.best.com/~herndon/scan/scaninfo.html	John Herndon's Scanner Files (Freq Lists)
http://www.zianet.com/kc5kto/scanner.html	KC5KTO's Scanner Page (Loads of freqs)
http://www.li.net/~j4dice/scanli.html	Long Island Area Scanning Resources (Northeastern U.S. freqs. and info.)
http://www.ntia.doc.gov/osmhome/nebbia.html	NTIA Spectrum 137 - 10000 MHz
http://www.artscipub.com/repmain.html	Open Amateur Radio Repeaters
http://www.vivanet.com/~duffingl/index.html	Polar Orbital Weather Satellite Homepage
http://www.aa.net/~lutz/radio/fun.htm	Radio Frequency Page (General Freqs)
http://wvlink.mpl.com/users/hunt_m/scan.html	Scanner Frequencies (General Freqs)
http://www.cdc.net/~deliryum/freq.html	Scanner Frequencies (General Freqs)
http://www.peak.org/~bueno/pla010.html	Scanner Frequencies (General Freqs)
http://www.gim.net/cp/1995/0720/src/article004.html	Scanners
http://www.ibmpcug.co.uk/~g4sgi/	Simon Collings; G4SGI Home Page
http://www.bssc.edu.au/student/walde.h/spook1.htm	Spook's Spot Links to Scanner/Radio Info.
http://exo.com/~rbarron/	Stupid Scanner Tricks
http://members.aol.com/wwhitby/trs.html	Trunked Radio Systems Users Page - Info on trunked systems throughout the country.
http://www.ozemail.com.au/~air491/	Victoria Police Website (unofficial) (Aus)
http://www.acs.oakland.edu/~ncschult/boatfreq.htm	Great Lakes Vessel Passage Frequency Page (Maritime Frequencies)
http://www.acs.oakland.edu/~ncschult/boatfreq.htm	Great Lakes Vessel Passage Frequency Page (Maritime Frequencies)
http://www.acs.oakland.edu/~ncschult/boatfreq.htm	Great Lakes Vessel Passage Frequency Page (Maritime Frequencies)
http://www.acs.oakland.edu/~ncschult/boatfreq.htm	Great Lakes Vessel Passage Frequency Page (Maritime Frequencies)
http://www.acs.oakland.edu/~ncschult/boatfreq.htm	Boat Watching With a VHF Scanner
http://www.navcen.uscg.mil/marcomms/vhf.htm	VHF Marine Radio Channels and Freqs
http://ourworld.compuserve.com/homepages/bcheek/	Scanner Mods by Bill Cheek
ftp://ftp.qrz.com/mods	Scanner Modifications (qrz.com)
gopher://hamster.business.uwo.ca:70/11/.mods	Scanner Modifications (uwo.ca)
http://lonestar.texas.net/~jac//scanner/	Some Scanner Modifications
http://homepage.dave-world.net/~woody99/index.htm	Woody's Pro-2004/5/6 Page
http://home.ptd.net/~pro2006	The Monitoring Post Pro-2006 Info Page
http://www.shellab.com/~yngwie/pro51.html	Pro-51 Keyboard Trickery
http://www.winstoncuponline.com/frequencies/busch.html	1996 Busch Series Scanner Frequencies
http://www.winstoncuponline.com/scanner.html	1996 Winston Cup Series Scanner Freqs

http://www2.msstate.edu/~rls3/bgnscan	Busch Grand National Scanner Freqs
http://www.fred.net/downsouth/truck.htm	Busch Super Truck Scanner Freq Listing
http://www.uniden.com/scanner/scanner.htm	Indy & NASCAR Frequencies
http://users1.ee.net/rp/indyfrq.htm	IndyCar Frequency Page
http://www.indy.net/~sperkins/scanner.html	IndyCar Scanner Frequencies
http://emporium.turnpike.net/~caj26287/racelink/	IndyCar Scanner Frequencies
http://www.speedcenter.com/garage/scanner.html	IndyCar Scanner Frequencies
http://www.geocities.com/MotorCity/1865/freq.htm	IndyCar Scanner Frequencies
http://student-www.eng.hawaii.edu/carina/	NASCAR Winston Cup Scanner Freqs
http://www.intercom.net/user/jemoore/freq.htm	NASCAR Winston Cup Scanner Freqs
http://www2.msstate.edu/~rls3/wcscan	NASCAR Winston Cup Scanner Freqs
http://www.newwave.net/~rbarker/nscrfreq.txt	NASCAR Winston Cup Scanner Freqs
http://mall.turnpike.net/~caj26287/racelink/scannerf.htm	NASCAR Winston Cup Scanner Freqs
http://www.execpc.com/~jnorton/nascarfr.html	Winston Cup Scanner Frequencies
http://www.fred.net/downsouth/winston.htm	Winston Cup Scanner Frequencies
http://www.aho.com/amtrak/freq.html	Amtrak Radio Frequencies for Scanners
http://bjr.acf.nyu.edu/railinfo/scanning/	Bill's Current Scanner Setup (New York City Area Railroad freqs.)
http://129.93.226.137/rr/rrfreq.htm#cn	Canadian National Rail
http://www.webcom.com/paine/FreqLA.html	Rails LA Scanner Freqs Los Angeles Area
http://www.loa.com/lasalle/frequencies.html	Trains of New England Railroad Freqs
http://zippy.cso.uiuc.edu:8080/~roma/rr-freqs.html	U.S. Railroad Frequencies
http://www.getnet.com/~n7qyk/	Roy Schahrer's (N7QYK) Home Page (Scanner Specifications; etc.)
http://www.flatoday.com/space/visit/scan.htm	Kennedy Space Center Shuttle Frequencies
http://www.intrnet.net/~gwilson/frequency/govt.html	Federal Government Frequencies
http://www.scimitar.com/revolution/by_topic/science/	Federal Scanner Frequencies
http://www.underground.org/publications/informatik/	U.S. Secret Service Radio
http://www.winradio.net.au	Rosetta Laboratories; support of WinRadio
http://www.winradio.com	US site for WinRadio
http://members.aol.com/j4dice/frequencies/travel/	Yosemite National Park Frequencies

Index

173